1

© July 2012

Cover painting : Cees Deen

Author: Pro. dr; Egbert K. Duursma
Member Academia Europaea
Em. Professor University of Groningen
Em. Director Neth. Inst. Sea Res.
and Delta Inst. Hydrobiology of the KNAW Nl.

Corresponding Address:
© E.K.Duursma
320 Av du Sémaphore
06190 Roquebrune-Cap Martin, France
duursma@orange.fr

The ZWIKKER CODE

Clearing mankind's
Indoctrination

Egbert Duursma

*'Help!... help!' shouted three frightened trembling schoolgirls,
when they saw arriving a small tele-commanded helicopter
which halted above the mob of aggressive boys that
surrounded them. A supersonic sound made all holding their
hands at their heads, while the boys screamed: 'Flee! Get
out of here! A heli-zwikker!' They became disoriented,
looked at each other and disappeared in all directions.
The girls cried clearly relieved. 'It worked! We're safe!'*

1

*Who feels himself the centre of the earth ...? Exact ...!
And who preserves the State ...? Exact again ...!*

Washington, 2061

Tim Peasley scowled surprised at the screen in front of him. 'I'll be darned. Why is this Bolotnikov in New York? Usually Presidents make plenty of fuss beforehand. Does the strange news from Uzbekistan have anything to do with his visit?'

He pushed a button on his desk and shouted: 'Harry..., come here..!!!'

Harry Snowdon, his secretary, blew in.

'Can I help you Sir?'

'I want all the news about Uzbekistan! And about the visit last week of President Gennady Raskadov to Moscow. Something strange is going on in Russia! Send me Mike Spone.'

'Anything else, Sir?'

'Shut up. Do as I ask you. Don't waste my time.'

'O.K. Sir,' replied Harry, gliding around the table towards the door. He was for long accustomed to the bullying attitude of his boss and had learned not to worry about it.

Peasley was the head of the United States Security Service. Since international intelligence was assured by Interpol, falling under the United Nations, he had only federal tasks to worry about. He was not allowed to carry out international intelligence activities, a restriction he ignored. He kept a system of electronic spying and telephone control.

Most people were shocked when meeting him. Anything that could burst out of his jacket did so. Even his arms stuck out at an angle. As a bodybuilder at the university he was nicknamed "Mr. Muscle". His features betrayed little of his thoughts. He had no success with women and refrained from seeking their company. Once, being deeply in love, he had proposed to a female co-student. She had burst out in such a fit of helpless laughter, she could not stop for ages. This hurt him so profoundly that he vowed never to be laughed at again, never...! He finished his studies and made a lightning career. His work was his life and his subordinates knew he was always working and within easy reach. He expected people to carry out his wishes, and those who knew what was good for them invariably did so.

'Sir, you asked for me?'

'Wipe that smile off your face, Spone. This is not a social call. I want to know all telephone conversations over the last two weeks between the Russian President Bolotnikov and the Secretary General of the UN, Dolores Rodrigues Guerrero. Immediately!'

'But I'm still busy with the urgent task you gave me yesterday. Shall I finish that one first?'

'I said IMMEDIATELY! You know that word? It's plain English, isn't it? Why are you still standing there? Come back in five minutes and don't keep me waiting.'

In the corridor Spone passed the secretary's office. 'What's up with the boss, Harry? He is in one of his tempers.'

'Don't try to understand, son. Something's in the air, that's for sure. You'd better hurry if you want to keep your job.'

In five minutes sharp Spone returned with a small key, which Peasley grabbed from his hands. 'Get out Spone, I want to see it alone.'

A shocked Spone walked out mumbling: 'Not even a "thank you!". He's worse than ever.'

'What is worse than ever..?' asked a voice behind him.

'Oh, nothing, nothing..., really nothing.'

'That fellow is becoming very odd,' said the voice to himself, seeing Spone hurrying away.

There was only one message. It puzzled Peasley more than he would admit. He had a very quick and suspicious mind. He never trusted anybody but himself, an attitude which had helped him to survive several Presidents, in spite of their general aversion to his personality.

"Can I see you in the morning at ten? The subject code is Bol-56," he heard.

'Why did he phone in the middle of the night? Dolores might not have appreciated. This is hot stuff... But what can be so hot about consciousness and subconsciousness? It must have to do with the visit of Raskadov in Moscow the day before.'

'Harry, give me our contact person in Moscow. Quick.'

'Jonathan Hawk, Sir?'

'Don't ask questions, I know it is Hawk. Just do it.'

His man appeared on the screen within a few minutes, panting and looking annoyed. 'Who wants to see me in such a hurry? Oh..., sorry Sir.., I see... Oh! Excuse me.'

'Shut up. What is up there in Moscow with Raskadov and Bolotnikov?'

'Something strange, Sir.'

'I know that already, why do you think I call? Be brief!'

'I've only got the summary of the threetel[1] broadcast of Raskadov's visit to the Duma, Sir. Two minutes after he started his speech he suddenly stopped, looked very surprised and said, with a strange smile: "The rest of my speech does not matter. It is the usual bla-bla you are used to in this house ...". He greeted everybody and left the hall. His wife stumbled along behind him, also smiling strangely. And she is something like Socrates' wife. Always peering around to see whether Raskadov is honoured enough.'

'Never mind the Greeks. Tell me exactly what this man did before he obviously changed his mind.'

'Yes..., now you say it. He spoke just normally when he thanked the Speaker of the Duma for receiving him. Then he stopped and pushed with his right hand on something on the bench. That something he had in his briefcase when he arrived. What it was, I don't know, but....'

'Send me straight away that summary, I know enough,' interrupted Peasley and cut off the connection.

'Gennady Raskadov became a lamb after he pressed on that thing.... Strange! I think Bolotnikov played a trick with him..., something like hypnosis. Then he went to tell the Secretary General of the UN about it. They often confide in each other; too much to my liking... Let me see what Internet tells me about Uzbekistan.'

He typed the name of the Republic. The latest press communications and individual reports had suddenly changed. Initially they spoke of their admiration for their great leader, Raskadov. At least until last week, the date of Raskadov's visit to Moscow. Then the messages became confused.

'These people have gone stone crazy. Saying how nice everything is, and without mentioning why. Raskadov is a

---

[1] (3-dimensional TV)

darling! My foot! That bold frog, who bullied everybody. Bizarre! Ah, here is a note from the Orthodox Church. They also embrace the doctrine of Mohammed. Who ever heard such nonsense?'

He remained looking at the screen, but he could make no sense out of it. 'As much as those people were indoctrinated, so free of mind they seem to be now. That's perhaps the clue..! That something on the bench suddenly changed the behaviour of Raskadov *and*, in the same time, that of his people! There surely must be a missing link. I'd better find out quick!'

His telephone rang and the face of President John Smith looked at him from the small screen. 'Tim, why haven't I been informed about the sudden visit of Bolotnikov to the UN? It's your job to keep track of visits of foreign Presidents, isn't it? Anyway, I already know why he came and what subjects he discussed with Dolores Guerrero. I'll send you the information and I expect you tomorrow at 7 p.m. at the White House. Be there, with some good explanation!' The face of the President zoomed away before Peasley even could reply.

'Good Lord! Call me a duck!' he exclaimed angrily. 'What the hell is going on? Harry, pass on immediately the video from the White House as soon as it arrives. Don't look at it..! This is classified..!'

In his office, Harry shook his head. This was really a day.

Peasley started the telephone recording which he obtained from Spone. He saw the impatient face of an elderly woman, probably a secretary, who was trying to get through. She let the call ring for several minutes. Then she dialled another number.

'Probably a code,' Peasley thought. 'I ll check that later.'

"Finally, Mrs. Guerrero, can you hear me and can you see me?" he heard, "I am Lena Dubrovkina, secretary of the

Russian President." "Yes Lena, I see you and hear you, too. What's the matter..? Why are you waking me up in the middle of the night?" "Mrs. Guerrero, would you please turn on your computer before I connect you to the President?"

Peasley heard a click. The screen became hazy and after a few seconds the image zoomed in on the Russian President.

"Are you there, Dolores?" "Yes, Victor. Is that all you want to know? Can I go to sleep again?" "Don't push the button, Dolores, please... I must see you urgently. Listen, and don't be upset of being disturbed at this late hour. I perfectly know you've had an exhausting week. But believe me, I must talk to you first thing in the morning."

Peasley watched the ensuing discussion, but nothing of additional interest followed. 'This Bolotnikov was really worried,' he mumbled to himself. 'Let me see what he told Dolores the next morning.'

In the mean time the video of the President arrived. It tuned in on Bolotnikov, who spoke in short sentences, mastering his English perfectly, as was required of all heads of State. Dolores replied frequently and was asked to place a disc in her threetel video.

'Ah! The same I received from Moscow. Interesting! I now see the change in attitude of that frog Raskadov.' Peasley let the images advance more rapidly, until he heard Bolotnikov saying: "We really had a narrow escape. A zwikker was stolen from our centre by a scientist who was probably one of Raskadov's agents. We suppose that Raskadov intended to indoctrinate the Duma during his visit with this equipment in the same way as he did with his own population. In other words becoming the boss of Russia. So we arranged an exchange with a prepared zwikker of our own. Just look at this video."

A three-dimensional picture showed the arrival of Raskadov at the airport. A solid woman with large shoes followed him with. She carried a suitcase. Raskadov shook hands with the Speaker of the Duma in the VIP room. They let him pass the control post, while his wife had to deliver the hand luggage for control. "Here we exchanged the zwikker in Raskadov's suitcase." Raskadov's wife investigated directly after the control the case, but obviously all was all right. "Fortunately their zwikker was indeed a stolen zwikker, otherwise we should have been obliged to take other measures of which we were not able to oversee the consequences."

"But Victor, what shocks me most, is what you told me earlier this morning about that new invention.... We will be able to manipulate a massive part of the world's population. Man thus becomes a plaything of technology?"

Peasley stopped the video for a moment. 'What does Dolores mean with "... you told me earlier this morning"...? They haven't recorded everything! Let me see whether they discuss this topic again. It must have to do with Raskadov.' Peasley restarted the video.

"That's correct." He heard Bolotnikov saying.

"In case you are not too familiar with philosophy and psychology, I will give you a short introduction."

'Oh boy, that man remains always a Professor. Why not be brief and come to the point.' Peasley sighed. Instinct told him that he was about to be bored. 'I have to suffer through all his rhetoric.'

"Since man became a rational being. ......

"I have read some books about it. Common primordial consciousness for example."

"Indeed. I always compare it to computers. Life is programmed by a number of "software" packages. The human brain acts as "hardware" with a large degree of freedom. The software is continuously adjusted through evolution."

"What your theory does not resolve, is the beginning of it all?" said Dolores.

"You're right."

"Are you trying to tell me that we are now capable of measuring this "software package" with an instrument which you call zwikker? But that is fantastic! You've certainly made a great scientific leap forward."

"Indeed, but there is one big problem. We'll be able to measure the contact between people through their consciousness and subconsciousness. And..., both can be influenced. Do you realise what that means? Many positive things, but also negative... We'll finally understand why people want to live together, but ....," Victor hesitated, "what will really happen when our common consciousness is directed by an instrument? If this goes wrong, we'll create monsters like Frankenstein."

"Perhaps, but you can also cure religious and political fanatics, who often incite terrorism and start wars. You don't mean we should refrain from further research into this invention, do you? I presume you're going to give me the answer. That's why you came to see me, isn't it?"

"The first question is whether individual States and the United Nations can handle this problem. Once you have it, the zwikker has to be kept under control. It's not easy, as you saw with Uzbekistan. Or..., should we leave it solely in the hands of scientists...? That's too risky. They could be tempted to use it at the expense of others. Indoctrination is not uncommon in our world, but not by a machine." Dolores sighed: "I suppose you have a solution, don't you? Is this man with you the inventor?"

Peasley stopped the video to think things over. 'These people became quite philosophical. Bloody nonsense. And those religions, it's all the same, they want to become immortal and their rites and bureaucracies do the rest. That's the trouble when you mix religion and philosophy with politics. But what about the invention?'

Switching it on again, he saw Victor introducing a handsome man of about thirty years old. His features were different from the Slavic type. He was slim, well dresses and spoke with a slight accent.

"Yuri Kaspadov, at your service," he said politely. "It is a great honour and pleasure to meet you and to explain our new invention. In case I will go in detail too much, I ask you to stop me."

"It is a pleasure to know you, Mister Kaspadov....," Peasley let the video advance more rapidly...   "You   may   have gathered from my accent, that I am not an original Russian. I was born in Siberia, from a generation that was displaced in the twentieth century..."

'Deported ....,' mumbled Peasley.

"... I obtained a position in the Russian Space Centre to carry out research on gravitation. I used an instrument which produced something very unexpected. I've brought it with me to show you."

Dolores looked first at him and then around in the room, but she saw only a small suitcase in the corner. "Does that contain your invention?"

"Yes Ma'am. I will complete my story.....,"

'I can skip all the bla, bla. This fellow is trying too hard to impress her.'

The suitcase contained an ordinary video threetel recorder and a silvery box. Kaspadov took a mini video disk from his pocket and placed in the recorder.

"The screen shows a water polo match. Half of the supporters were in favour of, let's say club A, the other half supported the other club, we'll call it B. The supporters were wearing the colours of their own club."

"My equipment was installed in one of the press cabins. Its sensor was directed towards the supporters group A." The screen showed a graph of peaks and dips. "This view differs slightly on one place, when the sensor is focused on the B supporters. This is due to the difference in devotion to their clubs."

Peasley could indeed see a slight change in the peaks in one section, when the sensor was focused on the other supporters.

"The invention is capable of calculating and preparing a counter pattern of this particular section. When transmitted, it will wipe out the original section. The result is partial loss of consciousness and the group of supporters looses its devotion to their club."

The video showed again the polo match with shots of the public. The noise was tremendous and slogans were being shouted from one side to the other. In the left corner of the screen appeared the text *compensation A transmitted*. The supporters with the red shirts, probably group A, stopped shouting. One man, who stood with his hands on his mouth, howled something like 'Davái, ái, ái, ái, ...!' and halted with open mouth in 'Dáv ...'. He and the others looked at one another, totally surprised, as if thinking: "... what the hell are we doing here?" Some of them removed their shirts and left the stand. The same happened with the second group when the text *compensation B transmitted* appeared. The stands emptied slowly, although the match was still going on.

"Now the second demonstration." Kaspadov took the silver box and a small black sensor from the suitcase. He connected the box to an electric point and to the threetel.

"You see that vase of flowers on the table. When I focus the sensor on the vase, I make the gravitation visible on the screen of the threetel. The pattern is much simpler than the one you saw in the former demonstration. The next action is more or less the same as for the water polo match. I keep the sensor focused on the vase, calculate the compensation pattern and transmit it to the vase."

Peasley watched attentively. Nothing happened, until Victor blew gently in the direction of the vase. Then, to Peasley's surprise, the vase moved out beyond the edge of the table and remained there, suspended in the air. Kaspadov turned the first button slightly and the vase slowly descended to the floor.

'I've seen enough,' mumbled Peasley, let me pass on quickly the images. 'They really have invented something. We must get our hands on it as quickly as possible. Let's see whether this inventor explains how it works?'

"..... is the compensation-effect permanent?"

"Probably. The removed consciousness does not regenerate itself. It may build up again, but only by contact with other people who still have this kind of consciousness. It is different for the vase. The moment the compensation of gravitation is shut down, gravitation again takes "possession" of the pot. Shutting down can also be done slowly, as you saw during the demonstration."

The video ended there. Peasley stared at the empty screen for a long time.

'There must be more, otherwise President Smith wouldn't have called me in tomorrow. Let's see if anyone phoned him from the UN?' Peasley typed his opening codes and found only one call, with the Secretary-General.

"John, Victor has arrived this morning and he wants to see you tomorrow. The Vice-President of the European Union, who is accidentally in New York, and some representatives from Australia, South-America, Africa and Asia, will also be present. The topic will be sent to your computer. I expect you tomorrow morning at 10 o'clock. This will give you ample time to read the message and change your agenda." Her attitude was almost North-American, but the President failed to notice it.

"Dolores, stop. I am in the middle of a meeting on the annual budget, which will last all day. Tomorrow I have to be in California!"

"Just read my message first and then judge for yourself. Take an expert with you and don't talk about it with anybody. You have a Vice-President, haven't you?"

'That must be the message I've received already. They're certainly going to discuss something hot and the President wants my opinion tomorrow afternoon. I have to be ready.' He dialled the number of his agent in the UN on his wrist-telephone. After a few seconds his man, Theodore Place replied.

'Here Peasley. Can you hear me?'

'Yes Sir, loud and clear.'

'Do you have the Secretary-General's schedule for today?'

'Partly, Sir. She received the Russian President and had several other persons in her office.'

'I want to know immediately the names of those people on my computer, and all the info you can get about them. If their discussions with the Secretary-General are recorded, send a full copy. You know how to brake in?'

'No problem, Sir. I'll see to it immediately.'

Peasley broke off. 'At least one man who reacts professionally.'

It took Theodore Place only twenty minutes to inform his boss in Washington.

'There were four people with the Secretary General, of which you can find their antecedents on your screen. A Chinese, an Australian, an African, and a Brazilian. Mrs. Guerrero talked with all four of them together and they will attend a meeting tomorrow morning together with President

Smith, President Bolotnikov and the Vice-President of the European Union, Julien De Beaufort. That's all the info I could get.'

'Thanks Flace. Keep your eyes and ears open,' snapped Peasley.

'This must be the reason why Smith wants to talk to me tomorrow evening. It seems an odd mixture of people. I don't like it. Dolores Guerrero is handling things again her own way. That Chilean woman acts often too highhandedness. This will give trouble and I am sure that De Beaufort will certainly start with: "Why wasn't I informed long beforehand..?" This man is the stuffiest shirt I ever saw at such a high position.'

'Let me examine what really happened in New York. Harry..., I need again Spone!!' shouted Peasley.

Two minutes later Spone came rushing in, panting heavily.

'Yes Mr. Peasley, what can I do for you?'

'You bugged the residence of Mrs. Guerrero two years ago, did you?'

'Indeed Sir. We regularly control it.'

'Then I want a recording of each word spoken in the house, including the telephone, from this morning until they go asleep.'

'We can also take images, Sir. Would you like them too?'

'Of course. Put your best men on it and secure only those discussions of value.'

'You mean about the Russian President, Sir?'

'What do you think otherwise, about their sex life or their cooking? Send everything to my home tonight.'

'Aye, aye, Sir, you will get them as soon as possible,' and Spone withdraw.

It was indeed very late when Peasley received the image and voice recordings. He viewed them from his large hoverbed. They came through loud and clear, so much that he felt himself an intruder. He saw Dolores Guerrero returning home very tired. Jorge, her husband observed her anxiously.

"What happened darling? You look so pale. Did Victor give you a lot of troubles?"

"Ooh, that's expressed weakly. Just let me come home first. I've never before been so shocked. This may be something bigger than I can handle."

"Yes, I imagine from your call this afternoon. ......"

'Bloody Spone, he hasn't recorded that call.'

"...... relax first. I prepared something special for you. A receipt from my mother. One of her good things. You don't have a zwikker with you? Funny name for that invention."

"Oh gosh, no..! But I saw it working. I felt a strange tension in my head, something I never felt before. That thing is dangerous. Jorge..., I will tell you all in sequence. Some part you know already, but by telling this again..? Perhaps it will help me recovering..."

"Just as you like, darling, I will listen."

"Also when I repeat myself..?"

"Also...."

She started with the arrival of Victor and the demonstrations given by the inventor......

"My God, this Victor has saddled you up with a real problem. And tomorrow morning you have to come up with a plan? How thus that bloody thing works?"

"Along the same principle as an anti-sound box, where sound waves are compensated by phase-shifted echo-waves."

"I like those anti-sound boxes. It's each time funny when that stupid dog of our neighbours barks without a sound with an astonishing stare on his face..."

"The equipment contains an ultra thin gyroscope, closed up in a vacuum box, which turns so rapid that the metal molecules of the disc reach almost a plasma phase, which is usually only possible at very high temperatures. The axis of the gyroscope is sensitive to the "waves" of human consciousness and of the forces of gravitation. All happens in the emptiness of the atoms and molecules. You remember the theory of a chemist 40 years ago in which he claims that the volume of this emptiness is thousand billion times larger than the volume of the atomic nucleus and that therefore the largest creation of the universe is the fact that atoms don't collapse. Negative electrons are not absorbed by the positive nucleus for periods of billion of years. This emptiness seems to be in some way or another "alive", thus acting with our spirit.

The invention is a great step forwards in science. It becomes possible to cure criminals from their complexes and transform them in rational human beings. But who guarantees that the effect will be permanent? Jorge, it makes me so tired..."

"Anyway, it is not your task to play the almighty. *What the wind brings along carries the wind off!* After you phoned me this afternoon, I ....."

'Again that missing call. They must have an additional protective code which we haven't cracked yet...!' Peasley stopped the recording for a moment. He was tired too. 'Just let me see the rest,' he mumbled.

"I have made up my mind about it." he heard Jorge saying.

'Aha, perhaps something interesting from that weak-headed pop-eyed poet. A fellow who lives on his wife's salary. I hope he will not muzzle things up...'

"I suggest you allow a controled and supervised development of the equipment, zwikker you said, wasn't it. Parallel to this development you must set up a research to detect them so it can be destroyed when necessary. I am not a technologist, but when such a thing can send out signals or waves, some other thing must be able to detect them and destroy it." Jorge showed such a wise face that Dolores could not help laughing.

"There is something more I was thinking about. You said that Victor mentioned indoctrination, either through education, schooling or religion. This gives a growing-up child a feeling of an own identity. We are mammals of a social kind and probably our instinct is not enough. Wrong and nonsense naturally. These children just build up a common consciousness, which is practical, that's for sure. Anyway it inspired me to a poem. You want to hear it?"

"Please, should I?"

"Just this one, it will help you, Dolores."

"O.K., I'll listen."

The world flees a way aloft,
our fate fades away!
We think, we work.., we're really keen?
Is knowing perhaps too much?

Are we alone in spirit,
or can we draw from a source,
which lastingly rustles around?
Ah.., if I only knew...

Wherein tails my identity...,
when I just think like other's delusions..
Or gives this a rightly sense ..,
to my integrated existence?

"Jorge, you're a genius. I know now what I will suggest tomorrow. These ideas give me hope again. The equipment can help us humans to understand ourselves. And that might have many advantages. The question of security is second, but should be treated at the same time."

'Ooh..!' sighed Peasley. 'Should I go on?'

He saw them turning on the threetel and heard a discussion on the world population of 15 billion.

"I go to bed, are you coming?"

"Yes.., yes. I am waiting a call from Natasha."

Peasley knew about them. Natasha was their daughter in San Francisco and their son Robert worked in Montreal. There was nothing special in the call and Peasley heard Jorge afterwards mumbling: "It would be interesting when the zwikker could determine our all ance. It becomes enormously thrilling when the zwikker discovers fields of science of which we never dreamt. Finally we may be able to detect why or how an object "knows" that it is attracted by the earth..., and the so called "ether" becomes visible. And why....!"

Where is the bridge between body and mind?
In the small infinity of matter?
In the emptiness of an atom?
Conscious now, or always been unconscious?

Instructions rush back and forwards,
a leg moves.
The mind commands,
the body reacts, as in a delusion.

Externally radiates a clear wave,
measurable as a signal.
Is our spirit therefore omnipresent,
and is consciousness its derivative?

What remains when I die?
Spirit alone, or something more?
But where does it stays?
and by who is it inherited?

Perhaps are we able to measure some of it,
an emptiness, in and outside,
timeless and always filled.
Enough for knowing?

'Stupid poet,' sneered Peasley and stopped the recordings.

2

Is freedom that what another will tolerate?

Washington

'Alfred, Walcom and Tim, we encounter a world crisis if we are not very careful with the information which I received this morning at the UN.' President John Smith paced up and down in the oval room of the White House. 'Tim, you should have informed me weeks ago about the strange things happening in Uzbekistan. I only learned this morning that its President Genady Raskodov managed to posses a so-called zwikker. He had it probably stolen from the Moscow centre of Space Research. Raskodov indoctrinated his population completely and was one step away from doing so with the Russian Federation.

Imagine what would have happened. Fortunately the Russians set up a trap to catch his zwikker. They exchanged it at the airport for another one, loaded with a compensation programme. When Raskodov pushed the button of this zwikker during his address to the Duma, he wiped out the consciousness of himself, some members of the Duma and of the people in Uzbekistan who followed the direct broadcast. Those in the Duma, who had plugged in for simulation translation, were not affected. The world escaped a real unprecedented disaster.'

Tim Peasley wanted to interrupt, but the President went on: 'We came this morning in New York to an even unprecedented decision, which will st r up a lot of dust both in the UN as in many a government. We, the Secretary General, Vice-President De Beaufort of the European Union, a chosen group of people originating from different continents and myself came to the following decision:

❖ The zwikker will not be destroyed. With the present level of technology and science, it can be re-invented within a few years time.

❖ The remote detection and destruction of the zwikker will be studied by an international team in Moscow.

❖ A possible therapeutical application of the zwikker will be studied in the US, also with an international team.'

'But Mister President, you and the UN have practically broken with all the rules of agreed governing!' shouted Walcom Sermon. 'How ever can I explain this in the Senate? They will kill me before I can open the session.'

'I know, I know.... The same will happen with the Secretary General, believe me. I must say I'm impressed by her attitude. She is not afraid to face the General Assembly and the press. Perhaps, because she is convinced that we should handle quick. And what counts also, she does not mind to loose her job. I'm encouraged by her. With a problem as large as this one, anything above the level of standard rules seems to be allowed. You cannot protect the world with given electoral promises and existing bureaucracy. Not only you, but also Julien De Beaufort will suffer. He started with: "why wasn't I informed before, so I could have requested council." Even our own security service has not warned me in advance!' He turned his head to Peasley.

'What is all this nonsense?' reacted the Senate Speaker. 'Aren't Bolotnikov and his reckless research workers cheating you?' He was clearly suspicious.

'Unfortunately not. Our problem is that we don't possess this new invention. We can only participate with medical and technical research within the framework of the UN. However, nobody can prevent us to develop a detection apparatus by our self. We will be criticized from all sides if we don't, also by the different religions. For that reason I will speak tomorrow with the Papal Nuncio whether he can arrange a discussion between the religions.'

He continued: 'We made agreements and we have to keep them, but let it be clear to you, we will take immediately protecting measures. We need to build this detection system which can trace zwikkers as soon as they "transmit". I have understood that the instrument produces a kind of radiation which can be measured on distance.'

'But, Mr. President,' spoke Gregory, 'before developing a detection apparatus, we should have a working zwikker. Did Bolotnikov leave such a thing behind?'

'No, he didn't.'

'Isn't it easier to ask for the construction diagrams and build the thing here?' interrupted Peasley. If we must wait for someone in the US to invent the zwikker it will be too late.'

He thought pragmatically and disliked detours. At a danger one should intervene immediately and all those international political arrangements would only lead to confusion. An outbreak of an epidemic had to be nipped in the bud, was his motto.

'Perhaps there is a solution. I will provide the UN with a couple of specialists for the technical research in Moscow in addition to the medical research in New York.'

'Aren't we complicating the matter, Sir. Let me catch a zwikker and the US will have the result you desire.' Peasley seemed more enormous than he was already. His bald head glowed. He had always to master himself in the company of politicians which he found weaklings who paid more attention to their own position then to the interest of the US.

'I support your proposal Mr. President,' spoke the Senate Speaker, ignoring Peasley's remark. 'That prevents difficulties with the Senate and Foreign affairs.'

'We will free a large budget from our reserves and involve our best people and universities. The US Security Service shall take care of security,' concluded the President

Peasley looked completely flat with no expression on his face, only his nuke muscles rolled. 'A plan full of holes,' he thought. The world should not be governed by politicians who make concessions to NGO's, churches, synagogues and mosques. This will give us only extra work and our service is blamed when the plan fails. Perhaps I can find someone who can pass me the drawings of the zwikker.'

'And no espionage,' spoke Smith, looking at Peasley. 'If something happens, I want to learn it first from you.'

'You say so, Sir,' Peasley murmured to himself, keeping a poker face.

'What concerns the discussions with the world religions, Tim, I want you to find a humanist whose antecedents are excellent so he or she can join constructively in a religious conversation.'

Peasley nodded and thought: 'That's at least one good proposal. I surely find one that keeps me fully informed.'

The President asked if there were further observations and concluded with: 'Thank you for your presence?'

When all had left, he pressed the intercom for its secretary Peter Scott. 'Are there urgent messages, Peter?'

'Yes, Mister President, the Prime Minister of Canada is on the line?'

A baffled Prime Minister of Canada, Bill McIntrow appeared on the screen. His nick name was "the flame" due to his thick red Scottish hair.

'Are you there John? I learned of that zwikker. As I understand you are more or less caught by our charming Secretary General and subsequently you have, by looking in her eyes, forgotten me.'

'How did you know?'

'Don't worry, we have our sources too. Will you act in some way or another?'

'Sure, but within the context of the UN. Be sure you will get all the information through them. And in spite of her eyes, it is serious this time. I expect you will support us.'

'When things develop as you expect, I suppose I have to, but inform me when things escalate, do you?'

3

For some believers god is safely far away. But suppose he's just close-by?

Washington

'Monsignor Altaldi, Papal Nuncio to the US,' announced the secretary.

'Welcome, Monsignor Altaldi, I appreciate very much you could come at so short notice. I have to discuss with you a subject which is probably of serous concern to the various religions in the world. But before I start, have you followed the news yesterday?'

'Mister President, my answer is negative. I was at a number of ecclesiastical meetings and did not have enough time for reading the papers.'

'Well, then I suggest you read first this report and press bulletins. It takes no longer than twenty minutes. You will be faster informed this way. I will leave for ten minutes, so you will not be disturbed.'

The President left his office and went to Joan, who kept her coffee pause. It was one of the moments of the day when nobody could disturb her.

'He, John, also coffee? You seem somewhat stressed. What is the matter?'

'Pooh, I have the Papal Nuncio with me and he reads at the moment the bulletins concerning the zwikker. I'm curious to know his reaction.'

'I also, but don't worry. The Catholic Church exist already for more than two thousand years and this leaves its tracks.' Joan had always a quieting influence on him. It was not his responsibility how the Catholic Church or any other religion would react.

'You're right. That coffee is delicious, really. Let me enjoy it for just ten minutes. You can arrange we get coffee or tea in my office?'

'Of course, just ring.'

John went up again and found the Nuncio pacing up and down in his office. He turned around to him. 'Is this all true? Would an instrument be capable of measuring a faith in the Lord? And not to speak of indoctrination and compensation as if a religion can thus be classified? Mister President, this will completely upset the various religions. Is it true that in Uzbekistan the Orthodox Church embraces the Islam?'

'I don't know more than you just read Monsignor. Did not Christ leave a certain freedom for own initiative, and do not have all religions their own responsibilities?'

'Indeed, and I understand now why you called for me. If I assess this well you want to know how we should proceed. Or putting it differently, how we can protect humanity against an indoctrination by means of an apparatus '

'You are not frightened that the apparatus will be used to measure the subconsciousness of individuals or groups?'

'Not at first sight. What would a religion be worth if it depends on a spectrum on a screen? On the contrary, the zwikker shows that there is more between matter and mind then we know, or rather that mind and matter are different than we knew so far. Although a faith always remains a struggle, there is more than we can observe. If one can deepen our knowledge by this apparatus, the religions may benefit from it. Perhaps a time comes that we know what all religions have exactly in common, so we focus less on the differences.'

Smith had not expected such an answer. There was more in this man he had thought. No wonder he had mounted to this position.

'Do you think your colleagues will share this conception? And would the Vatican approve a discussion with other religions on the zwikker? Such a discussion may be very worthwhile. I can arrange a place for a meeting and share travel and accommodation expenses.' Like always the President was practical and came rapidly with a proposal.

'Perhaps this is still somewhat opportune. I understand that today the whole world is informed and that reactions are bound to come. I will contact as soon as possible the Holy Father and present your proposal to him. That's all I can promise for the moment.'

'Thank you. I like to include also humanists. We should not be restrictive.'

The Nuncio nodded. In the last century a better understanding had grown between Christianity, other religions and humanism, although the number of sects had increased. The idea hit him that with a zwikker he could exhort sects. He placed the idea aside in the box of "sinful thoughts", so that he would have something to confess in Rome. That became more and more difficult at his age.

'Mister President, I suppose the Holy Father will ask me to come to Rome for a conversation. I thank you for your information and confidence in me to treat this matter further.'

The Nuncio withdrew and was guided by the secretary to his vehicle.

Vatican

Monsignor Altaldi strode slowly over the Via Colombo, the main street which led to the Vatican. The view on the Saint Peter Basilica was impressive. The Nuncio was worried. He had not succeeded to get the Pope on the line and the conversation with Cardinal Rogerius, who was responsible for foreign affairs, had not been very successful. The Nuncio understood completely that the Pope seldomly held telephonic conversations. The clerics distrusted this means, in spite of the fact that the telephone companies guaranteed complete privacy.

Cardinal Rogerius had obviously not seen any press bulletins on the zwikker and had reacted very cryptically. 'A conversation with the Holy Father should be requested one month in advance, you know that my dear friend. Where we would be, when everyone can pop in with the Holy Father, even a Nuncio.'

The Nuncio had answered: 'Monsignor, you can learn from the recent press bulletins that there is an emergency concerning an invention that considers us directly. The President of the US has requested me to inform the Holy Father personally.'

The Cardinal had replied that his task was to judge the importance of such a request. On arrival in Rome he found a message that an appointment had been made for ten o'clock in the morning the day after.

The Nuncio sighed. Although a supersonic flight gave no particular problems, it was for him still four o'clock in the night. In spite of one day extra rest he felt sleepy. He had remained long awake thinking thwarted by the hierarchical structure of his own church. In other situations he found this structure excellent. It had preserved the Catholic Church through the centuries.

Cardinal Rogerius was known as an influential man. He became cardinal at a young age and made his entrance in the Vatican not long afterwards. He managed worldly matters and international relations. His influence on the Holy Father was large, but more because he took much work from the Pope hands, then by stipulating strategies.

The Cardinal received the Nuncio in his own office behind the Saint Peter Basilica.

'You should have been clearer in your telephone conversation. I have only yesterday understood from the press bulletins the commotion about the zwikker and discussed it already with the Holy Father. He showed himself very concerned and he wants to be informed about the request of

President Smith. If you had passed this message to me, I could have prepared already the answer. Now we will lose much time.'

The Nuncio bent slightly. He found the reproach undeserved, but resisted strongly the inclination to give a sharp answer. The subject was far more important than a personal friction, how understandable it may be.

'When can I speak with the Holy Father?'

'In half an hour. We have enough time to discuss the decisions which must be taken. Do you have an opinion how the church should react?'

'Of course, I had one day longer and I have reached to the conclusion that the church must take the initiative and start consultations with other religions.'

'You are not afraid that this so-called zwikker weakens the faith in our church?'

'Could be, but will not.'

'This I don't understand. When the zwikker removes our faith-consciousness, nothing remains, isn't it? It would become possible to indoctrinate the faith-consciousness of a Protestant. Think of that!'

'Perhaps, and therefore studies are required.'

'Then we must prohibit these studies!'

'And just that has no sense. Governments are also frightened for the zwikker, but once prohibited or destroyed, it can be invented elsewhere again. Just like the dragon of Saint Gregory with many heads.'

'This dragon had only one head. You should know that.'

'The course of our conversation, Monsignor, makes clear that we have to do with a serious phenomenon and I propose to postpone our discussion after we have heard the opinion of the Holy Father.'

'You are right,' the Cardinal answered. Although rapidly excited and being attached to his position in front of a lower in rank, he was quick of apprehension. He admired in silence the

Nuncio who had expressed himself with more wisdom than he had shown.

'Let us go to the residence of the Pope. That takes us exactly ten minutes. You have the documents with you?'

'No. I think that a conversation has more value, in particular when it is registered. That is possible I suppose?'

'I will ask the secretary to arrange that.'

After having taken the lift, they were allowed into the suite of the Pope. He welcomed them warmly. Refreshments had been put ready. The Nuncio asked for coffee and the cardinal took a juice. The Pope himself took nothing.

'Monsignor Altaldi, I appreciate it particular that the President of the US is asking me, or rather us, to initiate a consult with the world religions. But before we come to that point, I must receive some details from you. Let me start with a question. Would that so-called zwikker be able to measure the difference between a non-consecrated and a consecrated host? It seems to me that a negative answer is of no importance, because it would prove that the zwikker cannot measure this matter. But if the answer is positive, then we are obliged to take the phenomenon zwikker very seriously. Please take your time for an answer. You also Monsignor Rogerius, but I want particularly the answer of Monsignor Altaldi.'

The Nuncio sighed. In order to think objectively he sipped quietly his coffee and he tried to conceive the many tasks of the Holy Father. Pope Paul VIII, who originated from Central America, was of Indian birth.

The Nuncio thought: 'He looks more like an Inca then many of his compatriots. It would be interesting to know whether he has a similar subconsciousness as the former Inca's.'

Thinking about that helped him to formulate his answer: 'One is not convinced whether the field of consciousness measured by the zwikker is similar with the field which exists between matter, or between matter and spirit. With the same apparatus one can measure both fields, however, but there is still no proof that the fields are the same. The chance exists however that there is a relation. The church has never denied that matter contains forces which strengthen our faith. Relics for example. We know also that people feel something special during a pilgrimage or at places of former catastrophes... In other words, matter can hold a field of a former event. My answer concerning the hosts is that in principle such a difference could be measured. In case the actual zwikker fails to measure the difference, an improved version might do it in future. In any case we must take it into account. For me personally it does not matter. My faith does not require a technical affirmative.'

The Pope nodded. 'What do you think, Monsignor Rogerius?'

'I wonder what the ordinary believers would think of it. In any case we must prepare council directives in which they can find support. We cannot leave it to everyone how he or she should interpret this phenomenon.' The cardinal had spoken with clear voice and looked critically to the Nuncio who had not mentioned this point.

'The church does not have the name to react rapidly upon events,' spoke the Pope. 'Whether this is a good behaviour, is for the moment an open question. The phenomenon, about which we speak here, concerns all humanity, that's clear. Whatever religion one supports, or if one does not believe at all, we must reach with each other to an opinion and if possible action. If not, soon someone will use the zwikker for personal aims. Then the world awaits a new kind of slavery, and this time the slaves are in agreement with their destiny. I have therefore decided to follow the suggestion of President Smith and bring together the world religions for a thorough conversation, with the aim of reaching a common opinion. Even when this does not succeed completely, we have at least enriched ourselves with a discussion. A council can wait for the moment. I request you, Monsignor Altaldi, to organise a conference between the religions and I give you thereby the rank of special Papal Delegate. Cardinal Benedictus will assist you. He speaks many languages, knows the other religions and works rapidly.'

'Where do you want me, Holy Father to take my residence and where do you think this conference should be held? Jerusalem for example? President Smith offered his help. He can contribute also in travel and accommodation expenses.'

'Personally have I nothing against Jerusalem, but perhaps the other religions do so. We must prevent that we mix state's interests and religion. I leave the choice to your judgement. I have appreciated your information and I thank Cardinal Rogerius for its attention in this matter. If you, Monsignor Altaldi are again in Rome, I expect an interim report. In case you need to reach me urgently I give you, at high exception, my phone number. There are only twelve cardinals who have to this number. You will be the thirteenth.'

'Let this number bring you luck too,' he smiled. 'I will give you, Monsignor Rogerius, tomorrow a pastoral letter for all churches. I will also ask the World Council of Churches to distribute this letter for further distribution, thus also Protestant churches are informed of it. I hope this work will simplify the task of Monsignor Altaldi.'

He requested his guests to kneel with him and prayed: 'Lord, you gave us a new trial which makes it possible to destroy mankind mentally, but perhaps may bring us closer to you. We pray you for support to reach good decisions. Give us the inspiration and wisdom to surmount this trial together with all other religions and non-believers, knowing that nothing on earth happens without your will. In the name of the Father, the Son and the Holy Spirit, Amen.'

They raised and bade farewell to the Holy Father. Cardinal Rogerius accompanied the Nuncio after having made an appointment for the following day with Cardinal Benedictus. 'You know him, do you?'

The Nuncio nodded. He was delighted to cooperate with Cardinal Benedictus. Where Cardinal Rogerius was the very image of Cardinal Richelieu, Cardinal Benedictus was literally and figuratively a frank person, who alone by his smile took people for himself. Someone who met him could not think differently than that the religion was a sociable matter.'

He walked relieved over the Saint Peter's Square. The idea that a lot of centuries looked down on him did him good.

'But oof, to act as a prophet will require all my attention,' he murmured.

4

Sometimes laymen may succeed where trained people fail.

New York

The same day it was buzzing in the General Assembly of the UN. It lasted some time before the press bulletins and the message of the Secretary-General dawned on all delegates. Twenty-two resolutions had been prepared. There was one which seriously blamed the Secretary-General for her obstinate behaviour.

Some delegations had misunderstood the problem and this led to all kinds of confusion.

'He, who has a zwikker, has an enormous military preponderance,' said a representative from a Western European country in the lobby. 'With this small box you can raise a tank of the ground. Interesting.'

'You misread the report,' retorted another. 'That box would need the electricity of a nuclear power plant for that job. You must read more carefully.'

'The box can let you think that it would be possible,' joked a third one.

'I find that we should support our Secretary-General,' said someone from South-America. 'We should all vote against those twenty-two resolutions, so at least she can go ahead.'

'Then you vote against my resolution, which just proposes this,' called a Dutchman.

The conversations became very alive and continued until the bell went. The General Assembly filled itself again and the Secretary-General came in.

She sorted her papers and spoke: 'Dear delegates. I have done something which deviates from the standard UN procedures. My excuse is only that I've done this with the best intentions for which we stand all together.'

'We encounter today a vital episode in our existence. An apparatus has been developed, zwikker it is called, which can do bad, but also good things. I hope with the help of your approval, we can rapidly go at work and get the Zwikker under control. I will show you now a video demonstration which says more than I can express in words. I thank you for your attention.'

It remained very quiet in the Assembly. The delegates had obviously been taken by surprise, expecting a much longer speech. Using this reaction, Dolores gave a sign to the technician in one of the glass boxes in the room. Light weakened and the delegations were looking to the three-dimensional images of the water polo game and the experiment with vase with flowers.

When the video stopped, a loud hullabaloo went on. The President hammered for silence and adjourned the meeting for an hour.

Of the twenty-two resolutions six remained and after exact seventy minutes, the polls included, the meeting was over. The Secretary-General had got the support of most of the delegations, although in the adopted resolutions a number of restrictions had been built in. She had these already imposed by herself, so that it had no further consequence. A couple of delegation members congratulated her and wished her much success with this important task. Most countries were prepared to provide members for the Supervision Commissions.

Dolores returned to her office and dialled Professor Bolotnikov. She got him after ten minutes on the line.

'Oh Victor, you sleep with a bonnet?' she roared of laughter.

Victor withdrew it rapidly and muttered: 'Is this your revenge for last week? To disturb me in the middle of the night? What is the news?'

'The General Assembly gave its blessing, isn't that enough? We can now go ahead and you should be the first to know. Now you can sleep really quiet, don't you think? Victor do not comb your hair, it is splendid this way.' She pressed the "off" button to grant him no reword.

Washington

During the time the General Assembly discussed the zwikker, Peasley called one of his undercover employees, Ilja Travenkov. This Frenchman of Russian origin was already for a number of years a *mole* for his agency. He had a position at the travel office of the Western European Union in Strasbourg. By this way he recorded all travels of the officials. Once per month he forwarded by e-mail an encoded overview to Peasley.

For the first time in his career he had got the request to come to Washington. He had taken a week holiday and took the first plane to Washington. Now he just walked in the city of which he had heard so much. The building of US Security Service disappointed him. He had expected a skyscraper in modern style, but not a house of a couple of hundred years old.

'Why they hide themselves while there is no reason to do so? Everyone in the world knows them?'

Travenkov presented himself at the gatekeeper and showed his chip card for a screen. The gatekeeper called the secretary of Peasley who requested him to send him up. Travenkov passed a wide hall with a serial of huge portraits of the former directors. The painters had tried to give them strong features, but sometimes this had resulted in crooked grimaces. What they all had in common was that they looked tightly at the visitor and from the beginning until the end of the hall. Travenkov checked whether their looks were still targeted on him at the end of the hall.

The secretary left Travenkov in the large office and closed the door behind him. Peasley had never met Travenkov and as an exception he stood up to welcome his visitor. This usually never happened twice. Since the floor of the office was slightly lower than that of his desk, he had a psychological advantage to them. His impressive shoulders and the enormous chair did the rest.

Travenkov, however, had no experience with such intentions. He always tried to be at the service to his customers. His character radiated kindness. The travel customers frequently told him about their travel and brought back strange currencies, which he collected. Travenkov was middle fifty, carried thick glasses and was shabby dressed. As a widower he had not remarried and his only son lived in Russia.

This fact had brought Peasley to select Travenkov to steal the plans of the zwikker. By means of his son, Peasley could influence him.

'You still speak fluently Russian?'

'Sure and I maintain it by reading Russian books and follow the Russian threetel transmissions.'

'You are in our service,' said Peasley superfluously and waited the affirmative. 'I have your Curriculum Vitae here on my desk.'

Travenkov stood up, caught the document and looked at it.

'Indeed Mister Peasley, it is correct. I have, however, done nothing else for you then sending the travel overviews, and none was lacking. In fact what do you do with my reports?'

Peasley ignored the question. 'I have another task for you. You must take a month leave. You explain that you want to visit your son because of happy family circumstances. Your new-born grandson is named after you and you want to assist at his baptism. In Moscow you have to steal a zwikker or the plans for a zwikker. This is most important for the US. In any way you must avoid to be connected with us. You will leave tonight for Moscow, after you have arranged with Strasbourg that you extend your holiday with three weeks. You have sufficient reserve holidays, they will easily understand that.'

'What are my work conditions? And what help you give me during my espionage activities?'

'We have chosen you because you are unknown with Interpol and the Russian security service. You knew for a long time that you as a *mole* could be called for a task outside the usual pattern. If you are caught, we will tell that we don't know you. You get an amount of money that cannot be traced. It stands, as it happens, already for some years on your account. So far you've never known the number of your account. The bank however knows it and that you regularly used the interest. We did that in your name. The credit card lies in front of you and you can see that even your signature is correct. Since the bank does not know you personally, but has your photograph, there is absolutely no problem.'

'And if I do not want to carry out this job?'

'I do not expect this of you. You love your son and grandson and you do not want something to happen with them. On the contrary, now you are rich, you can help them.'

He thought further: 'Why I have to work with such a little man. What a pity we cannot send trained people. The Russians would smell them straight after their arrival. As a matter of fact Travenkov should have known the consequences of being employed by a foreign security service.'

'You have, Sir, selected a non-experienced person for espionage. Are you convinced I can execute this task?' Travenkov hoped clearly to be able to escape.

'You don't think I considered this already? You are indeed the less bad solution. You must take the task and not ask any further questions. You can withdraw now, study the documents and learn these by hart. I know you have a photographic memory. The documents will be destroyed afterwards. When you leave from here no trace is left.'

Travenkov stood up. All kind of thoughts came up, but the expression on Peasley's face held him back to reply. He did not like Peasley and hoped he would never see him again. With a short good-by he left the office and went to the secretary.

The secretary brought him to a kind of cell on the second floor with small window exposed to a garden. 'There are beautiful asters,' he thought. He put himself to reading and for an hour he memorised the pages three times. The zwikker interested him intensively. Should he be able to influence Peasley? He suppressed this idea of revenge. If he obtained the plans, there could be other interested parties than the US Security Service. But Peasley must have thought of that too and probably people would watch him and intervene at the smallest deviation from his task.

The voyage to his beloved Russia attracted him. After checking all memorised data, he requested a call with the head of his service in Strasbourg. To his surprise it was absolutely no problem to extend his holiday. The chief was very pleasant and understood his request and asked: 'Is the weather nice in Washington?'

'Very beautiful,' replied Travenkov, then realising that he had not informed him on his destination. 'I visited today a number of historical places when my son called me. He became recently father and wanted me to assist at the baptism of my grandson. I will suspend my plans for America and set off tonight for Moscow.'

Travenkov was proud on himself that he had found so rapidly an answer. Or would they watch him too? So far he had never noticed any interest.

'Better take it into account,' he thought, 'I might have a number of European security agents on my heels. If Interpol joins them it s a complete herd. Coo, it starts well. Did Peasley tell the truth? If I had a zwikker, would I be able to check that?'

He sighed. 'What would Jana have said of this?' He missed her. She died now ten years ago and he had never looked for a new relation. 'She would have been able to help me, because a travelling couple is less conspcuous. At least what concerns espionage,' he thought. Jana was a strong woman who had a thundering laugh. She had become somewhat deaf and as a result she spoke fairly loud.

Travenkov rang as agreed the secretary. The secretary took the documents and handed him the travel documents and the credit card.

On the street the traffic moved still ahead and Travenkov had the unreal feeling that he dreamed. Was he now the selected master pawn who had to steal the world most dangerous invention? Or at least the plans? And why was this, actually? On his way to Washington he had read that the zwikker fell under the UN and that only in New York and in Moscow research should be carried out. Why then this complication to penetrate a centre in Russia? An American member of the Commission of Supervision could just "lend" the plans? Or would Peasley want a zwikker for himself? That fat pig veiled more behind his poker face then he had shown.

Travenkov was intrigued by the zwikker problem and as an ordinary person he was flattered to get a world task. It was paid well too. He was now millionaire and could soon live on his investments. Then he could at last travel to those places where his clients frequently told him about. He had never been really jealous, but his wish for travelling had been always there.

He halted a taxi which brought him to his hotel. There he dialled the international number to block his new credit card, wrapped up his luggage and went for lunch. His plane departed only at midnight so he had enough time to enjoy Washington.

In his office Peasley sneered when he saw that Travenkov had blocked his credit card.

'That fellow is less dumb than he looks,' he thought satisfied.

He called Harry and ordered: 'Look for a humanist who is on our payroll?'

'Just a minute, Sir. I will put the result on your screen.'

A moment later a list of persons with their photographs appeared on his screen.

'What a bunch of softies,' he muttered in himself. 'They seem as hypocritical as religious people who want to look holy. 'That person there, Brill Dowson is perhaps something. She seems shrew. Her C.V. seems acceptable when it is slightly polished up.'

This took him no longer than ten minutes and e-mailed the file to the President. The President glanced at is shortly and ordered his secretary to send the name to Monsignor Altaldi.

Strasbourg

The buzzer on the office of Vice-President Julien De Beaufort was humming. He got the tendency to turn the thing off. The last days had been hectic. Everyone had gone mad over him. The country representatives found that he had insisted too little in New York in obtaining more knowledge on the zwikker. One feared, and not wrongfully, that the US, China and Japan within a short time would have developed their own zwikker leaving the Union aside. In spite of the cooperation in Europe, national interests dominated still above common goals. The rich variation of cultures and languages had been, however, preserved, but in common matters leadership was often absent. The Union had developed a by far-reaching democratisation into an almost independently working bureaucracy. Even with a Vice-President with a more powerful character this would not have changed much.

De Beaufort pressed on the intercom. It was his secretary who communicated that the head of a department, under which fell the travel office, a Mr Jenke Holthus, wanted to see him urgently. De Beaufort asked for the reason for this request, but Holthus had refused this to the secretary. He only said it was very urgent.

The Vice-President hated conversations which had not been prepared by his counsellors. And this was just a head of a low-ranked service. Times seemed to have changed since he had returned from New York. Perhaps it was a subject with which he could recover some respect.

'O.K., let him in and give me his state of service.' These appeared on the screen at the same time with the knock on the door.

'Come in!'

A typical Scandinavian appeared. Fair, with a broad face, blue eyes and arms full of hair.

'Sit down Mister Holthus, just a minute.'

This left him some time to look through the antecedents of its visitor. Holthus seem to be already for several years in the organisation. He had a military training and had been appointed as a head of a department which also supervised the Union computer systems. He had no function at the security service of the Union.

'Why do you want speak me so urgently? You should know that I have little time for sudden visitors.'

'The US Security Service, Sir.'

'What do you have to do with them? And what does that concerns me?'

'In our service is a certain Ilja Travenkov. He works in our travel office. Travenkov has taken a holiday to Washington, but rang after one day that he wanted to extend his leave for three weeks to visit his son in Russia. For the baptism of his grandson. You don't find that strange?'

'Mr Holthus, if this is all you want to tell me, you can go. I do not interfere with holidays of my staff. If a holiday gives problems for your service, you must solve that yourself. And would you be so kind to leave my office, I have urgent things to do.' De Beaufort took a document and went studying this.

Holthus did not move.

'Haven't you gone yet?' said the Vice-President after some time. 'Do you want me to put a bed in here? What time do you like to be called in the morning?' Sarcastic, of course. Bitter, undoubtedly. But there are times when a man may legitimately be sarcastic and bitter.

'Mr Vice-President, do you really think this was all? The number from where Travenkov rang was that of the US Security Service. Travenkov is a so-called *mole* of the USSS! And then suddenly this behaviour. Even a less malignant someone then I can see that something is strange. He must steal the zwikker!'

The Vice-President jumped from his chair as if he was bitten. That zwikker had caused already enough difficulties to him and now this. 'What do you mean, Mr. Holthus and explain me exactly how you come up with this fantastic tale. You can read in the press that the zwikker falls under the UN and that the US is cooperating entirely. They participate in a research in New York and Moscow, so why should they want to steal it? Explain that to me!'

'To start with the beginning, Travenkov sends monthly an overview of the travels of our staff to the head of US Security Service, Peasley. I cannot conclude differently than that he is a spy. Because nothing further happened and he never travels, he must be called a *mole*. Travenkov is of an inconspicuous character, widower and speaks fluently Russian. Recently there was a problem with our computer. I worked a complete night to solve the failure. Then I noticed a report from Travenkov to Peasley. The failure concerned the coding that guarantees the privacy of bulletins, so I could read this report.'

Holthus remained silent to let it penetrate to the Vice-President.

'Here you have a copy of that bulletin. I haven't shown it to anybody and Travenkov ignores that I have it.'

'What do you want to do with this knowledge, Mr Holthus? You should have contacted our security service. That is your duty. You have kept back serious knowledge!'

'But Mr. Vice-President, do you understand that you have the chance to unmask the espionage of Travenkov? Or as you want, if he has completed his task, to use it for our own aims?'

'This goes too far,' replied De Beaufort angrily. 'We do not spy and I don't want to be involved with espionage of others. Moreover it is possible that you are entirely wrong. I am of the opinion that you must directly go with your so-called presumption to our security service.'

Holthus, however, did not let himself intimidate and played his last card. He knew that the Vice-President was a super-bureaucrat, but even these have weak spots.

'If my presumption contains just a little truth, it does not harm to examine it furthermore. I have read from critical press bulletins which problems the world faces. You and I am the only persons which are informed of this presumption of espionage and the only thing I ask you is to authorise a travel to Russia and keep an eye on Travenkov. I will act as if I am accidentally on holidays if I meet him. I will ask no advance on my travel expenses. After success I expect of course a compensation.'

De Beaufort had to admit that this scheme could not directly affect his position.

'Well. I cannot say that you have my blessing, but I will not stop you. Only if you achieve results I want to see you again, if not, I consider that this meeting has not taken place.'

In his heart he admired people such as Holthus who took responsibilities and were running large risks. He stood up and gave to Holthus a hand.

'You know my position, I wish you success. Don't discuss this to anybody, I really mean anybody!'

'Thank you Sir, I will be absolutely silent. I hope to be able to report to you whether Travenkov's mission has succeeded or has failed.'

Holthus left the office. He returned to his department, arranged his replacement for a month and went home. He booked his travel on purpose by other means then through his own travel office.

At home he packed some luggage, left behind a note for his wife that he had to go unexpectedly on an official trip.

5

How does science work? With luck or with wisdom, or with both?

New York

Six weeks after the UN debate, a zwikker had been delivered to the Central Hospital of New York. A wing had been reserved and provided with a number of securities. Interpol had been involved in screening the research workers. The group was not yet on full strength. They had to promise confidentiality and report only to the UN.

In the laboratory sixteen research workers listened to Professor Sven Larson head of the group, who gave a statement on the planning of the studies. Professor Larson was a famous Norwegian psychologist on group behaviour. In spite of the fact that he stuttered slightly, the research workers listened attentively to him. In its recitation Professor Larson tried to avoid the name zwikker as much as possible. Marga Stalma from California, thought that here was a beautiful case to cure the stuttering Professor.

'You c-c-can read that I propose a number of exp-p-p-eriments. You must consider this of course as a beginning. You can make several proposals on the basis of your own sp-speciality, but I would prefer to dis-dis-c-cus first my pr-pr-proposals. You follow me?' Professor Larson looked around pleasantly and disarmingly. Nobody laughed or smiled, but a couple of them had got a red head.

'A p-point is that we provisionally can-cannot study other persons than ourselves. We are there-therefore at the same time research worker and pa-pa-patient. Pa-pa-patient is per-perhaps not the correct word, because nobody of us is ill and our co-co-collective consciousness does not fall under the relation doctor-patient. We will work as follows. For each type of research we will sp-sp-split in two groups, research workers and pa-patients. This goes in turns, so everybody is involved in both groups. Is-is-is there a question?'

Marga Stalma stood up and asked: 'Professor Larson, I find your proposal splendid, but I nevertheless would gladly see that our tests have no negative impact on our own consciousness. Large influence tests must provisionally be postponed until we really understand the zwikker.'

'Of course, dea-dear co-co-colleague, I thought this would be obvious. There-therefore I propose to be the first test per-person. Undoubtedly you have noted you that I stutter and I hope tha-that one of you can cure me with the zwi-zwi-zwik-k-ker. And not in such a way, that you all will leave this laboratory as stammerers.' He grinned disarmingly and there went a sigh of relief through the room. One scientist stood up and asked the Russian technician, Boris Klemarov, to explain the functioning of the apparatus.

'Ladies and gentlemen, the zwikker is able to measure a paranormal field that could only be felt by people with paranormal gifts. Hypnosis also falls in this category. The apparatus is able to detect on submicron level the field that exists between matter and mind. How it exactly works is not essential for you. You will notice when I start the zwikker that it transmits an increasingly higher tone which becomes subsequently supersonic. This is caused by a gyroscope which is rotating at such a high speed that it approaches the speed of individual electrons in an atom. The material of the disk is of a special alloy and is resistant against this high speed. There is no danger that the thing explodes. The disk weighs only a few milligrams. The only danger arises when the field which is transmitted by the gyroscope is thwarted from outside, but for that another apparatus is required.'

Someone interrupted: 'Mr Klemarov, these technical details are particularly interesting, but can you demonstrate the functioning of the apparatus?'

'O.K., you are right. When you press this button the computer does the rest. On the monitor a demonstration spectrum will appear. Please pay attention. You may feel some reaction in your head, but don't worry about that. The spectrum which I project now is one of a painter in modern art. It is rather confused. He paints completely uninhibited.'

'I can show you still one more,' he went further, 'but I cannot explain them as a technician. That will be your task.'

'Thank you very much for your statement,' replied a slim young person, called Bertram Slinhald. 'Can I make some proposals? We come from several places of the world. Our individual subconsciousness will differ in some aspects. However, as soon as we start working together, a new group consciousness will build up. Mr Klemarov should therefore first measure us individually, before it obtains some common subconsciousness.'

'That's quite possible and I will do that straight away.' Klemarov invited successively each person to the former roentgen room of the laboratory which was completely enveloped with lead and had a lead-glass window. The sensor of the zwikker hung twenty centimetres above the head of the test person. Klemarov claimed that lead had no influence on the radiation captured by the zwikker, but it reduced any possible risks. All the other scientists followed the zwikker screen intensively and sometimes sheers were heard when a derogatory spectrum peak appeared.

From the individual spectra, it became clear that right part, the most recent one, a small identical peak had appeared.

'You see a common type of consciousness is already arising here. Over a week, if I will repeat this test, I think that this part of the spectrum will have grown. The rest is the "fingerprint" of your individual consciousness.'

A vivid discussion arose. One asked whether genetic deviations, like with Mongolians could be treated. Others proposed to examine cases of mental deviation or obsessions which are a danger for the society.

Another proposed that each of them should undergo psychoanalysis and correlate that with the patterns observed in their spectrum as found by the zwikker. The two psychoanalyses specialists could carry out this analysis in duplicate.

At the end of the day Professor Larson requested the computer specialist Jean-Pierre Petit to evaluate the results statistically. Jean-Pierre Petit was almost two meters long, thin and slightly curved. 'He certainly bends himself over his work,' thought someone who regarded him critically. The others collected their papers and left the building.

The first thing Jean-Pierre did was arranging all spectra in a three-dimensional way on the screen, one behind each other. The sixteen spectra resembled a moon landscape, in which absolutely no structure could be discovered. He selected for this reason two spectra and placed these each behind the other. Now the picture improved. Although some picks and valleys were different, there were similarities where one was longer than the other. This brought Jean-Pierre to the idea that age of the test person might play a role. For this reason he conducted the commando to classify the spectra to age.

The result was surprising. Jean-Pierre leaped from his chair by hurting his knee to the table. What he saw on the screen was a chain of hills in a valley landscape. The spectra of older persons were longer (more wide) and sometimes the hillocks were somewhat higher. He stared at the picture in full delight of someone who has made an unexpected discovery.

He mumbled: 'This zwikker really works. But why do older people have a different picture than younger people? Will older people have a broader consciousness? And why are there *überhaupt* hills and valleys? Would the zwikker detect the components of the subconsciousness separately? If that's so, then the rightist peaks of the spectrum are also the most recent parts of the subconsciousness while the left ones concern the deep subconsciousness'.'

He looked at them once again. 'And how should we explain the broadening impact? Each peak broadens himself with age? Are the valleys "nothing" or do they have a meaning? If a valley presents "nothing", how is possible that "nothing" broadens as well?'

Jean-Pierre shook his head. 'We must first finish the psychoanalyses exercise,' he mumbled. 'Tomorrow is another day.'

He ensured that all data in the computer were safely stored, pressed on the "Off" button, caught his briefcase and left for the hotel. There he would find the other colleagues. Which impression would make this news?'

6

Simplicity and faith... In small lies hidden large.

Moscow

Ilja Travenkov had arrived in Moscow. His son Pjotr had been at the airport and had taken him directly to his home. The complete family was present in the small apartment. For his grandson, also called Ilja, it was all the same. He was only three months old and ignorant of the fact that he would be baptised. An appointment had been made in the old Abbey where the mother of Pjotr was a cleaning lady. To Ilja's surprise the Abbey was adjacent to the gravitation research centre. Peasley had known this, however. For this reason he had chosen Ilja after long hesitation. Since his faith in his *mole* was not very large, Peasley had charged Mike Spone with the same task. Spone had already arrived in Moscow.

The mother in law of Pjotr, Natasha Shelnakova, was of Ilja's age. She lived in Pjotr's apartment which was appropriate for taking care of the grandson when she had no service in the Abbey. She was long time ago divorced and contributed to the family budget. Sometimes the mother, Olga, was peevish for living so close to each other, but since Pjotr had a moderately paid job and she worked only for two days per week, the contribution of the grandmother was helpful.

The visit of father Ilja gave more problems and she saw already part of her few savings disappearing. He would remain even three weeks. Her husband Pjotr had insisted that he should be present at the baptism of the little Ilja and this she could hardly refuse. Moreover she found father Ilja a kind man and he seemed to amuse himself rather well with her mother. They picked up many old memories and there was a lot of laughter.

The baptism went according to the rite of the Russian Orthodox Church. The small Ilja underwent everything with interest until he came three times in contact with water that was not entirely of room temperature. He expressed his fear and dissatisfaction loudly with as result that a wet family left the church.

Outside the Abbey stood someone, who observed them with interest. It was Spone who followed Travenkov already for a couple days. 'If this is the man who has to steal the plans of the zwikker, I am a broad bean,' he muttered in himself. 'He rather has a good time with that granny Natasha. They went to a ballet show and an opera, and visited the Kremlin.'

Natasha had shown Ilja the Abbey where she worked. The Abbey dated from the seventeenth century and had beautiful frescos. Ilja found this splendid, since thus he might have the occasion to enter the building of the research centre. Peasley had told him where the research centre was, but he had not notified him that Natasha worked in the Abbey. 'That cunning fox,' thought Travenkov.

The Abbey was surrounded by walls and had several churches with their typical golden onion cupolas. The walls of the research building were painted white and the gardens were well maintained. Ilja and Natasha joined a flow of visitors who under the guidance of a guide visited the different churches. Since the explanation of the guide started to annoy them, Ilja asked Natasha whether she could show him her work area. Natasha took him through a number of corridors and they arrived at the building where the popes lived.

Here Ilja invited him for a cup tea in the canteen. What Ilja saw here, was a variegated collection of people in black and white.

'Who are those white-coats?'

'They are from the research centre. What they do is rather mysterious. They never speak about their work. As a matter of fact that would also be too difficult for me, because it seems to concern all electronic.'

'Do you come there sometimes?'

'Seldom, only when they need to extra cleaning help. They prefer to ask someone of us then someone from outside the Abbey. Moreover they know me by face.'

'And how do you enter?'

'Very simple, with my MasterCard. Those are appropriate for all doors. This is obligatory in case of fire. We work at night and if is there is a fire we must be able to escape. Look, it's this card.'

Ilja blushed to the root of his hair. 'Would it be this easy to go en steal the zwikker?' he thought. 'But I do not want to involve Natasha in my task. She is so nice and the work is too risky.'

'Is there something?' asked Natasha, looking at him over his fuming tea. 'You are sometimes so absent and looking in the distance. Just now you got it warm when I showed you that card. You have not only come here for the baptism of little Ilja, isn't it? And why are you followed by two different fellows?'

Ilja leaped up as a frog and was caught with his foot behind a chair leg.

'W-w-what fellows? You've seen them?'

'Oh, yes. You are here for another reason and it concerns this centre. Let us go from here and tell me everything. Perhaps I can help you.'

Ilja hopped behind Natasha. 'Am I then such a mug that one can immediately see that I am up to something mysterious?' he thought.

Outside Natasha said: 'Ilja, confess to me. What is it? I can be silent if I want and perhaps I can help you. If you had to sneak into that centre I know a possibility. You have thought of that when I showed you that card, isn't it?'

'I will tell you everything. Well almost for certain I am the only spy in the world who must steal the plans of so-called zwikker. I have to do that for the US Security Service and I cannot refuse. If I refuse, they will do something to Pjotr and the small Ilja. It's the head of the US Security Service, a certain Peasley, who ordered me to come to Washington. The baptism of our grandchild is just a cover up.'

'Where you've seen those fellows who followed me?' He had forgotten them a moment due to the emotions. 'Are they here?' He looked around as if they could be behind each tree.

'I will help you. Let us go home and make a good plan. Whether those fellows are here does not matter. I do not want any harm to be done to the little Ilja.' Natasha got something tough of the old Russia that had to overcome so many difficulties in history. 'We take a taxi.'

Since there was but one taxi on the square, their watchers could not follow them. Even if they had transport on their own account, the parking was outside the Abbey. Nevertheless Ilja looked several times over his shoulder, until Natasha seized his hand. 'Nothing will happen. And if so, we are together. Perhaps I must go with you to Strasbourg.'

Ilja blushed. Entirely bewildered he came home with Natasha and he was glad that Pjotr, Olga and the baby were not there.

The plan of Natasha was as simple as effective. The next day they went again to the Abbey. Natasha led Ilja by means of a number of corridors to a cellar which they had not visited the other day.

'Here above is the centre. We must wait in this cellar until eleven tonight. Until that time there are always people. If they leave, the automatic security is installed. For us this is no problem as long as we remain in this cellar which belongs to the centre. There is, however, talk of a new security system, but the sensors which were recently ordered have not yet been installed. There are now two guards instead of one. Those are present in the gatekeeper room. One does the rounds on the odd hours. We have enough time therefore and he cannot see or hear us.'

They waited until a quarter past eleven and went then ahead. Natasha had taken along soft gloves and a kind of mask. They almost looked like professional gangsters, except of the clothing of Natasha and the spectacles of Ilja. Ilja had no idea where he was going, but Natasha led him to a laboratory where many instruments were on the tables, some partially dismantled. In a connecting office stood some filing cabinets. Ilja tried to open these, but they were well closed. With his knowledge of computers Ilja succeeded to start one computer. He placed himself in front of the screen and searched for files related to the zwikker. Under the word zwikker he found nothing, but after many trials he found some kind of an entrance. The Russians had given the apparatus the name *Sovièst* which meant in English '*conscious*'.

A second problem was the access code. Ilja had experience with deciphering codes at his travel office at the Western European Union. After one half hour transpiring he succeeded to break the code. The plans for the construction of the zwikker came on the screen, but there was lacking information on the alloy of which the gyroscope was constructed. It lasted again eleven minutes before he also found this code. It took only a couple minutes to print two complete plans. He did not dare to copy it on a key.

When Ilja had finished, he removed the files from the computer and cleaned the key board. He had still enough time to look around for a zwikker, but obviously they had been stored somewhere else. Natasha and he withdrew themselves rapidly in the cellar. All together it had lasted almost one hour.

'Oof,' blew Natasha, when they had arrived there. 'That was quit thrilling. Something I've wanted to do once. You see it so often on the threetel, but in reality it is much more fun. Will the guard notice something in his next round? I don't think so. These two on service are not very bright. They rather like to eat and drink then pay attention at spies. What I said to you already Ilja, perhaps I must go now with you to Strasbourg. If I could choose, you are my man, I find you a real hero.'

'Is this a proposal?' Ilja had not yet recovered from the tension. He felt himself not in the least a hero.

'You could call it that. I always wanted to marry a frightened spy, but I never met one when I was a young little girl. No wonder my first marriage failed. How often I stood dreaming above my broom. She looked in the twilight younger than she was and Ilja felt vibrating a string of which he thought he didn't possess any longer. This string trembled first very softly, but gradually it became a concert.

'Perhaps it's not a bad idea. As a matter of fact we cannot do otherwise. The children have missed us tonight and I must defend you honour.'

'Folly..., come here, we must remain here for some hours before we can leave the cellar and I shiver of the tension. Let us shiver together.' They looked for a dry spot and although the floor was quiet hard, they fell soon asleep leaned against each other.

They awoke cold and stiff and it took Ilja some time to get his left leg moving again. They could unlock the door and walked through many corridors without encountering anyone. They came exactly on time for the tour of the first tourists and joined them. At the end they left this group and went quietly home.

They found an agitated Pjotr and Olga, who had waited the complete night for them. 'Where have you been, and why you haven't you warned us? We were just planning to call the police?'

Olga was faster than Pjotr: 'Or...., I understand... And that at your age''

'What do you mean,' asked Pjotr, who still thought of the police.

'See for yourself! Only a walk in the park, and that on a cold autumn night?'

'Exact,' spoke Natasha. 'Yes, that's what we have done. And we'll marry soon, isn't it Ilja?'

Ilja who was just cleaning his spectacles nodded a bit stupidly.

'Father, is that true?' called Pjotr.

Replacing his spectacles Ilja said: 'Yes, it's true, although I also know it just now. But.., since when have parents to give account to their children. I don't need to ask your permission, do I?'

'Yes, but you are both our parents? As a matter of fact, someone has rung for you. A certain Holthus if I have understood well. He is on holidays in Moscow and seems to know you.'

Ilja faded slightly and thought: 'That must be one of the peepers.' To Pjotr he spoke: 'Has he given his address or phone number?'

'Yes, you find it here on the notebook. He would appreciate when you call him back for an appointment.'

Olga, who found Pjotr's remark but a diversion, spoke: 'And when you marry, will you join father to Strasbourg?'

'Yes, my child, we agreed to that tonight and we have no reason to wait. We aren't that young any more.'

'And what concerns your problems of working and taking care of the baby, I can help.' Ilja interrupted. 'Tomorrow I will put an amount of money on Pjotr and the little Ilja, as a result of which you will have no financial worries for the next twenty years. Afterwards when we may have gone, you will inherit everything what we haven't spent. Since mothers dead I have lived simply and as a result, I have some savings.'

He ran to the house telephone and noted the number which had given Holthus. How could Holthus have known where he was?

'Oh yes' he thought, 'I've told him from Washington that I would be with my son, therefore it must have been easy to find Pjotr's address. But why does he want to speak with me? That is not clear to me. We've never been that friendly. Is he one of my prosecutors? And who is then the other?'

At twelve o'clock Travenkov rang the number which Holthus had given up. It was hotel Krymia. He was put through and a moment later he heard someone replying with a heavy Scandinavian accent. That could only be his boss Holthus.

'With Travenkov, I do speak with Mr Holthus?'

'Yes, you do. I would gladly invite you for diner tonight. I am occasionally in Moscow. Could you possibly come at five in the hotel Krymia. You know where hotel Krymia is, I suppose?'

'Yes, sure. I will be there, at five o'clock in the hall.'

'Until then,' and Holthus cut off the connection.

They had telephoned without picture. This was usual in hotels, unless one asked for it. Travenkov realised that he had not asked Holthus why wanted to see him.

He ran to Natasha. 'Have you a description of our spies? I mean you've seen them?'

'Not terribly clear. One was blond and the other some kind of Mexican. They did not work together. It was either one or the other who followed us. I've seen them only twice and then in a flash.'

'This blonde must be Holthus, I suppose. Then he is also after the zwikker. Why? Is the Union also interested or does he work on his own? We'll see tonight. He can't attack me in public and I have hidden the plans well.'

To recover from the experienced tensions Ilja laid down for a small nap. Pjotr and Olga were at work and the little Ilja did his afternoon doze. Natasha was shopping, so nobody would disturb him. Ilja dreamed of corridors, cellars with gates and monsters which leered on him. He had learned to be able to overcome his dreams, but this time he hardly succeeded. With a yell he awoke and realised it was three o'clock. Little Ilja who lay in a room beside him, started to cry.

Ilja senior had some problems to realise where he was, but then he remembered that he was in the home of his son and that his grandson was crying. He took the baby from the cradle and comforted him by rocking him back and forth. The smell he inhaled left no doubt he had to clean Ilja junior which he had never done before. Generally his spouse had done that in former days, but he had frequently looked on. He searched in the cupboard for a diaper and undressed Ilja junior. The boy found that splendid and very soon Ilja senior and junior were in a mood where no zwikker problems played a role. Olga had indicated where he could find a feeding-bottle and after bringing this on temperature, Ilja was feeding his grandson.

Thus Natasha found them at her return and there passed a wave of feelings through her. 'Ilja, I didn't know that you had it in you. My husband has in former days never done something like that with Olga.'

'I, honestly also not with Pjotr, but one can always start somewhere. It is a nice little boy and I will miss him when we are in Strasbourg.'

'Me too, we must see him often. At that age they grow like cabbage and before you realise he stands before you with a bass voice and girlfriend at his hand. It is nice that you will help Olga and Pjotr with money. They don't have an easy life, in particular with such a nosy parker as me over the floor. Do you have to go to that Holthus? Don't tell him I know everything. The less he knows, the better. I am curious what he wants.'

'I will go now. With the Metro it is approximately twenty minutes.' Ilja took his coat, set up a cap and left.

In the Metro he looked sharply around. There was no Mexican. There were, however, a number of youngsters who came obviously from school. They were in busy conversation and one of them mentioned the zwikker. He told that on the news was mentioned a burgling in the zwikker centre. An American, who resembled a Mexican, had been caught.

Travenkov jumped up. He wanted to hear more from these youngsters. But the conversation concerned already other things. The girls and boys had a lot of fun.

Ilja thought: 'If they only knew. Here I sit beside them, me a zwikker thief. Was that American or Mexican the other prosecutor?'

At the station Krymia he got off. The hotel had been once a modern hotel with much glass. It looked now somewhat neglected. Through the turn doors he came in the hall. He wanted to go to the counter but was stopped by a hand on his shoulder. With a jolt he twisted around and saw Holthus standing.

'Mr Holthus, I presume,' he said superfluous. 'How do you do?'

'Mr Travenkov, or let me call you Ilja, I have reserved a table in the bar where we can speak undisturbed. Is all well with your son and grandson? And with the lady with who I you saw these last days?'

Holthus certainly knew already too much. Travenkov nodded only he found it not correct to be called by his first name, but he followed Holthus without protest to the bar. This one ordered two vodkas on the rock and settled in one of the two chairs. He requested Travenkov to take the other chair in front of him.

'I know, Ilja, for what reason you are here. You are a *mole* of the American Security Service. By chance I came to know that from our computer system. When you suddenly went to Washington and still more unexpectedly continued your travel to Moscow, a candle started to burn for me. In Washington you visited the American Security Service, what I saw from the number when you rang. They ordered you to steal something in Moscow. So far, so good?'

Travenkov tried to keep a surprised expression on his face, but he succeeded badly.

'I see from your face that I'm right. I went to Vice-President De Beaufort and have his authorisation to examine your behaviour. Since I've seen you twice with that nice lady at the Abbey where that zwikker research centre is, I concluded that you had something to do there. You have not stolen a Zwikker, at least not when you left this morning the Abbey. What else, however? I therefore think that you have made a copy of the construction plans. The case of that lady was fuller this morning than yesterday evening then you went inside.'

Ilja took his spectacles and cleaned them. He was really at loss. Soon the Union and next the complete world would know about his robbery. What would happen with him? How could he marry Natasha when he was in the prison?

'Mr Holthus I don't know what you are talking about. Natasha, as is her name and I spent the complete night in the gardens of the Abbey. I have asked her to marry me and she has said yes. She is the mother of my daughter-in-law and lives with them. I find your conclusion very interesting, but it was on the afternoon news that a Mexican has been apprehended on suspicion of burgling the zwikker centre. Doesn't it seem a bit too accidental when two robberies take place at the same time in the same building? And since the robber is caught, how can that be me? My future spouse Natasha works there as a cleaning lady and in the beautiful autumn night we forgot the time. Haven't you been amorous?'

It was clear that Holthus was surprised. 'An American or Mexican, what do you mean? I have seen indeed someone who had also a large interest for your trips and that fellow looked like a Mexican. I have checked with the Union and they reported that he is, just like you, a member of the American Security Service. You don't deny that, isn't it? My presumption is that you are up to something which is closely followed by this American of Mexican origin.'

'I neither deny nor confirm that. You make presumptions without proofs. You assault me in my holiday and I have already explained that I will marry that nice lady. Is that not a sufficient reason for my behaviour?'

Holthus realised that he had to bring heavier artillery in proposition. 'Travenkov, you seriously endanger your position at your travel office. I have the proof that you did sent regularly travel overviews to the American Security Service and this directly to the head Tim Peasley. I have checked if you have an extra income, but so far this hasn't been the case. You are therefore a *mole* who gets a task assigned at a certain moment. You were chosen because you speak Russian and because the mother-in-law of your son works in the centre. I discovered your activities during a computer failure. You just had sent a month overview but the coding did not work. Afterwards I have paid attention to you and indeed you've sent each month a report. You were never thanked, because no bulletins returned. Don't you think that this is proof enough? In any case it is enough for the security service of the Union to imprison you. I assume that you don't wish to be a prisoner, certainly not since you want to start a new life. That's only possible if you hand over the construction plans of the zwikker to me. If you do this without resistance, nothing will happen to you, I can assure you. Moreover, as a result I become an accomplice. I will hand the plans over to Vice-President De Beaufort. You and I are pawns in a world game of forces. As I read, everyone is soon able to make a zwikker. Isn't it then a task of humanity if we ensure that our own states are not assaulted by criminals who catch such an apparatus? You hear that I have no self-interest. I have only my reasons with respect to the Union. You really risk nothing if you give those plans to me. You do, however, when Peasley gets them. Perhaps he acts outside the knowledge of President Smith. You have heard that I do nothing outside the Vice-President.'

Travenkov nodded. 'Well O.K. then. I certainly would like to preserve my position at the Union. I wish indeed to have a happy life with Natasha. Almost everything what you said, is true. Only Natasha has not worked that night, but smuggled me into the centre. She also knows that I have taken something. She thinks that this is for humanity, just like you formulated that. I was a *mole*, with emphasis on "was". I will not more be a *mole* when my task has expired. But if I provide no construction plans to Peasley, I get into troubles. He has threatened my family. I can give you therefore only a copy of the construction plans. The original ones I will keep to myself. I lose rather my job, then something might happen to my family.'

'I understand that. I accept your proposal. But I would gladly receive tonight that copy. I go along with you to the house of your son and wait outside.'

Holthus thought about the impact of leaving the original with Travenkov for himself. But if he informed De Beaufort, it was no longer his problem. De Beaufort could always warn President Smith to intercept the original.

They took a taxi and Holthus requested the driver to wait at the indicated address. Travenkov ran up the stairs and caught one copy from the cupboard and descended the staircase. Before the door he looked left and right for a Mexican, but he only saw Muscovites running from their work home.

Before he gave the papers to Holthus he requested a signed receipt on which he had written in Cyrillic "reception papers of "*Sovièst*". Holthus asked what it meant whereupon Travenkov explained that it only applied to both of them. Holthus checked with his dictionary notebook and signed. He wanted to leave now, put the papers in his suitcase and returned to the taxi. 'See you in Strasbourg,' he said.

Ilja was glad that nobody had been at home. Natasha was certainly walking with the baby and Olga and Pjotr were not yet home. For the moment he had escaped from a large danger.

He started the threetel to see more concerning the robbery. At the news of seven there was nothing. On the radio he just heard: '….. the Mexican American, who was arrested this morning at the research centre in the Abbey just outside Moscow, must have penetrated in the centre although the guards had seen nothing. Nothing was stolen, and the police had found nothing on him. The authorities are facing a riddle. The American has been taken for further questioning.

Travenkov sighed lighted. They had found nothing that could be related to him. The American would probably save himself.

What Ilja did not know was that in the centre they had indeed observed the use of a printer. Turned out documents were noted automatically. The capture of the American had conducted to a verification. They found quickly that the construction plans of the zwikker had been printed. But, where had that American left the plans? Did he have accomplices which no one had seen? The security service was involved and discussed this directly with Interpol without mentioning the use of the printer.

That afternoon, nine hours later than in Moscow, Peasley found a bulletin on his desk that Spone had been arrested in Moscow. Furthermore a burgling had been noticed in the zwikker centre, but nothing was missing. Peasley was notorious for its effectiveness. He called immediate his colleague of the security service in Moscow which he got on the line after ten minutes.

They looked at each other on the screen and Peasley spoke: 'Nicolai, what do I hear now. I serd one of my best men to test your security norms and you take him in custody. I've heard from the Interpol that your security system is not very good. For this reason the President of Uzbekistan could get hold of a zwikker. And I thought to help you by testing the system. Obviously it is better than the panic which you make of it.'

'Your tale sounds well, this time. You don't think I believe that. You do well to tell me exactly what is the matter with this man Spone. Burgling the centre is something else then testing a security system. If we had not arrested him, something serious could have happened. What was he looking for in the centre?'

'You know just as well that spying is forbidden. And that applies also to us. Whether you want or not, you must accept my tale as the truth. You will not hear from Spone a different story.'

'That I accept. But although nothing has been found on Spone, we have strong suspicion that you were up to something.'

'Something like that you don't say among friends. For that we need each other too much. You have possibly read that the world can become full of zwikkers if we do not stand together.'

'Tim, you've almost a holy wreath above your head and that stands you bad.' The two men looked at each other. Tim Peasley looked like a mixture between a bald bear and a pig. Nicolai Yesin resembled a lama, who looked down along his nose.

'Let Spone go. He has done you a service, but we don't ask thanks from you. Nevertheless you can count always on our support and god knows we may need each other soon. And to show my honesty I will send him to Uzbekistan where you have blundered so much.' Peasley continued to look at him as a self-complacency bear.

'Well, you win this time. Be aware however, if it is such as I suspect, we will get you.'

'Always at your service, my back is broader than yours, therefore I gladly take some of your problems. Just take better care of your security systems. Ha.., ha...' Peasley burst in a fat laughter and switched off the connection.

'Oof,' he sighed. 'That was a narrow escape. That stupid Spone. He had only to shadow Travenkov. For the moment I must wait what Travenkov performs. I cannot venture a new man, then they know for certain that there was something fishy.'

He rang his secretary and asked whether some urgent bulletins had come in. He, as it happened, had a golf appointment that he would not have missed for anything. Golf had become his large hobby when he had given up bodybuilding. He was a member of the most expensive club of Washington. He had enough money to spend. Otherwise there was only his home with a house keeper who restricted each conversation to "yes Mr" and "no Mr".

The secretary said that there were few urgent businesses. A phone call had come in from Moscow. The person had said that there was no hurry. He would take up contact tomorrow.

'How was his name?'

The secretary consulted the note book. 'Travenko or Travenpo, I couldn't understand it very well. He had a Russian accent similar to the man who was here a couple of weeks ago. You know, that fellow with the spectacles.'

Peasley controlled himself. 'When will he ring again?'

'Two PM our time, Sir.'

'And what else?' Peasley wanted under no circumstances to show his tension.

I will make a list and put that on your desk for tomorrow morning. Your taxi is waiting for quite some time to bring you to the golf field. Shall I inform the driver you are coming?'

'Yes, do that.' He interrupted the intercom, chose a sport-loving jacket from its cupboard, caught his golf clubs and left the building.

That afternoon he lost his two parties against a well-known bank director. This one was particularly delighted, but also astonished by the fact that Peasley took his loss so easy. That was new to him...

7

Try for once to be original within a rigid system! Wrong...!

Strasbourg

One week after Holthus had filched the zwikker plans from Travenkov he reappeared in Strasbourg at his office. After he had spoken with his employees over his holiday in Moscow, he rang the secretary of the Vice-President that he wanted to speak to him urgently. To the stupefaction of the secretary, the Vice-President replied that Holthus was welcome in ten minutes.

Holthus felt very content that he had been successful and he expected no less than applause of the Vice-President.

De Beaufort greeted him however somewhat cool. 'So, there you are again, Mr Holthus. I see it was all a storm in a glass of water.' He did not wait for the answer and continued: 'The Union is very grateful for your research. One can never know, isn't it?'

Holthus looked at the Vice-President with large stupefaction. 'But Sir, it's not like that. Here you have the construction plans of the zwikker!' He placed a map of papers on the desk of the Vice-President.

This one looked as if he saw a bomb lying on the desk. 'You do have...... the construction plans? But that's impossible! I have read that the zwikker centre in Moscow had been burgled, but that this burgling was only a cover up by an American spy, called Mike Spone, in order to test the security system. Nothing was taken away. And now you walk in my office with the plans in your hands? Do you realize what kind of an international scandal can be the result from this? Suppose it is announced that I have these plans? You should have considered this, Mr Holthus!'

Holthus reacted entirely bewildered. 'But..., you let me go yourself after Travenkov?'

'Not to steal the plans by yourself? How did you ever think of that?'

Holthus became gradually angry. There it happened again. If something good was achieved, it came on the account of the highest bosses, but if something went wrong it were always the lower officials who received the blame.

'Sir, I haven't stolen anything. Travenkov succeeded to steal the plans and it seems that nobody had noticed this theft. I've always kept him watched and I have seen with my own eyes that he has spent a night in the centre. Then I have made pressure on him and he confessed that he had two copies of the plans. One copy is in front of you, the other one Travenkov will hand over to Mr Peasley the head of the US Security Service. No one else knows about it. Travenkov knows better than that to tell Peasley that you too have a copy.'

De Beaufort, whose head became slowly very red, stood up. 'But that's even much more terrible. What do you think when it becomes known that I have the plans and moreover that I am informed of the fact that a certain Peasley in America has one too. Now, Mr Holthus, tell me that! I will be thrown out immediately on the street and I may be happy when I escape from a prison sentence by pleading that you have done this on your own initiative. And as a matter of fact, the Union has never commissioned you. The proof for that is that your travel and accommodation expenses are not paid by the Union.'

'As a further proof I take charge of these documents so you cannot do any harm with them. To prevent that your mistake will attract attention, I will not remove you from your position. On the other hand, when I hear you inform third parties on this matter, you will be punished. Mr Holthus, you can go.'

If ever someone had become waxy, it was Holthus. He left the office without saying a word. He was furious. After having been red, he became very pale, and the secretary who saw him going along, followed him concerned.

Arriving in his own office, Holthus took a firm glass of whisky from the bottle he had in reserve. He drank slowly.

'That scoundrel,' he cursed, 'that mean bastard.'

'He let me go on my own, supposing something profitable might happen, but now it becomes too risky I am the guilty party. I should never have left those plans on his desk. Now I am completely outsmarted. I cannot take anybody in confidence, they would declare me crazy. I should have made at least a copy. But even that I haven't done, I was too certain to get a compliment and a bonus. Now I've nothing.'

The whisky started to work and slowly he became quieter. 'I'll get him in due time,' he thought. 'I will make a detailed report and hide that carefully, so I've always a defence in reserve. I remember certain passages of the plans and I can mention those as a proof I had them in hands.'

Moscow

Travenkov was still in Moscow. To take Natasha to Strasbourg would be easy if they were a married couple. Since the administrative preparation lasted some days, he had to stay in Moscow. He had rung Peasley, but he was not available. Tomorrow evening at 14 hours Washington time, he would try again.

They would marry the day after tomorrow. Just simple, without much redundancy, only for the law. He wanted to deliver the construction plans of the zwikker as soon as possible and break with Peasley. Fortunately he had no longer seen Holthus and he hoped to be able to avoid him in Strasbourg.

The second telephone contact with Peasley came through rapidly. Travenkov had turned of the camera of the telephone so that Peasley's screen remained grey. He recognised however the voice of Travenkov and barked straight away: 'Are you there Travenkov. Is privacy guaranteed?'

'Yes Sir, I pressed the correct code.'

'And what do you want to say to me'

'In four days, Sir, at four o'clock in the afternoon at my travel office in Strasbourg. Your person must be able to legitimise himself, unless you come yourself.'

'Are you entirely crazy? You come here in my office and well immediately. Phone me as soon as you have landed in Washington.'

Peasley broke off the connection. When he gave a command, he never left time for objection. He was so used to this method that he did not expect a subordinate to disobey.

On the other side of the "line" Travenkov looked astonished at the little screen of his pulse telephone. He swallowed and said to Natasha: 'Do you mind to fly around the world to Strasbourg by means of Japan, San Francisco and Washington?'

'That seems marvellous to me! At last I become a globetrotter and then with such a fine fellow like you. You can arrange such a travel in short term?'

'No problem, I even get a discount because you will be my wife.'

Now he could travel quietly and nobody would follow them on this travel. Peasley had positively sent no other spy to Moscow, and in Strasbourg they were certainly waiting for him.

Thus it happened that Peasley received a week later a phone call from the airport that a certain Travenkov wanted to be picked up. He exploded almost. 'Letting me wait for one week and then calls me as if I'm his taxi driver. I will wash his ears when he is here. Let him take a taxi!'

It had been rather crowded on the road, and Travenkov arrived three quarters later. He had first dropped Natasha in the same hotel where he had stayed at the beginning of his holidays and had gone afterwards to the US Security Service. The building did not seem to have changed.

'Of course not,' thought Travenkov, 'I've been away only for three weeks. But it seems as much as thirty years.'

He walked through the corridor with the portraits and all eyes followed him again to the lift. Turning around he wanted to put out his tongue to them, but he controlled himself. 'What has happened to me,' he thought. 'The stay in Moscow has done me certainly well.'

'And Travenkov?'

'I wish you good morning Sir. There is a hole in your shoe. Unpleasant when it rains.'

Peasley quickly withdraw his feet from his desk.

'You've threatened me that something would happen with my family if I you did not carry out my task. You will receive from me what you requested, but I want first a written dismissal and a declaration signed by you and stamped by the Service that you will leave us entirely alone and in peace. In addition the name Travenkov should be removed from your computer files.'

Peasley looked astonished at the phenomenon Travenkov. A man with thick spectacles had carried out a task for the US Security Service and dictated his conditions!!

'Have you really the plans of the zwikker?'

'Yes Sir, and hereby I give you one part of them. The other part you will get as soon as you have fulfilled my conditions.' Travenkov gave him some papers and started to clean his spectacles. That this exercise heavily annoyed some people such as Peasley, escaped him.

'Would you please now erase all Travenkov files?'

The simplicity of this request baffled Peasley completely. He had rarely met such people. This very ordinary man had really thought of everything. He activated his computer and gave to command to remove all files related to Travenkov. A serial files came on the screen and the question was asked:

"Should these files really be erased". Peasley pushed OK and all disappeared from the screen.

Travenkov had seen that the correct procedure had been followed and said: 'Thank you very much Sir, would you please sign here?' Travenkov produced a paper from his briefcase and placed it on Peasley's desk. Peasley got his pen and signed.

'And here are the other parts of the plans, Sir.'

'He just simply had everything with him!' mumbled Peasley. 'And I, head of the largest and best security service of the world have been spellbound.' Peasley stared unbelievingly to Travenkov, but he had already stood up and said: 'I had made a second copy for myself, but it has been taken away by someone of the Western European Union. He knew I was a *mole* for you, and he had followed me to Moscow. He threatened that otherwise I would loose my job. He intended to give the copy to Vice-President De Beaufort, at least that was what he said. You see, it's no longer necessary to be a *mole* for you. It was a pleasure to have met you.'

Travenkov stretched out his hand which Peasley shook flabby and left the office. He greeted the secretary agreeably. Outside he felt he could dance; he had again his freedom.

It took Peasley five minutes to recover. This fellow had overwhelmed him entirely. He had even forgotten to ask how Travenkov had stolen the plans and whether they were genuine. He quickly looked at them and reached to the conclusion they were. As a result, he forgot Travenkov and proceeded to take action. He had to work rapidly, quicker than the Western European Union. He rang the head of his technical laboratory.

Until late in the evening they sat bent together over the drawings.

8

How true are scripts..., but if one asks nevertheless further?

Alberta

The call from Pope Paul VIII had not given the responses he had hoped. The disunion between the world religions existed still and most religions felt no need for a large conference of which the vertically structured Catholic Church was the organizer. The answers suggested a restricted and unofficial meeting of participants "à titre personnel" and that their opinion could not be explained as those of their religion or organisation. They were against a meeting in Jerusalem.

The proposal of the President the US to have the meeting in North America and to pay the accommodation expenses was accepted. The place which was chosen was in the State Alberta, Canada, close to an old Indian reservation. It was quiet there and one was not disturbed by the world press.

The arrival of the participants had had some delay, but on a Thursday morning in October a plenary session could start. Nuncio Altaldi was accepted as chairman. He knew bests most religions, whereas he spoke several languages, amongst which Arabic. This was useful if they would have problems with the final editing of their report. The archbishop of Canterbury, John Wiley, of the Anglican Church was chosen as rapporteur and secretary. Thus it was insured that the end document was in clear English.

'I welcome you all cordially,' started the Nuncio Altaldi. 'We may soon be able to check our faith in the Lord with a kind of box which can modify that faith or wipe it out. It is based on our subconsciousness. This is a very serious matter. It does not help if we hide ourselves behind religious texts. We can do that elsewhere but not here in this meeting. It is expected of us to discuss the core of the problem. Humanity is at stake. We stand on a crossroad. Our religious traditions can go in two directions: to be fully dependent of a technique, or to live in freedom with this technique, which can measure whether we are real or imaginary believers. I accept that you all are sincere as far as your faith or conceptions concern.'

He waited to let these words settle down and continued: 'In order to structure our discussions everyone in alphabetical will be asked to be the chairman for a half day or an evening. This will be sufficient for three days discussion. We start therefore with Cardinal Benedictus.'

Cardinal Benedictus got up. 'To go straight to the point, I propose we speak this morning on the impact which the zwikker can have on the existence of god. Who can I give the floor?'

The Muslin from Egypt, Ismael Mohammed Jami, gave a sign that he had no objection to be the first. Jami had been dressed traditionally. His fine-cut face resembled something of the old Bedouins on Arabic horses.

'I understand from your suggestion, dear colleague, that it does matter whether we believe in God or not. The question is: How is god involved in the mysterious field between spirit and matter, which can be measured with the zwikker? We must keep this sharply in mind.'

He remained silent as if he asked for inspiration. 'As I've understood, the zwikker measures a kind of field which is aroused by the human consciousness. Since Allah is believed to be omnipresent, it seems to me impossible that an apparatus can measure a divine consciousness. I think for this reason that the zwikker is no danger for Allah. Since people have been created by Allah, we rather observe with the zwikker a phenomenon which is only part of this creation.'

'This has not been proved of course,' entered Alex Whitewater the Presbyterian vicar. 'The problem of us is that we search for solutions of phenomena which are external of the limited possibilities of our brain. Most religions have been created since people in their frailty did not understand life. We believe, however, that scripts of Prophets and the Messiah helped. Since we don't have a Prophet who occupied himself with a technique, we have no guiding principle how to approach the zwikker. The zwikker remains therefore a matter on which we must come to a conclusion by ourselves. Whether God wants to test humanity with something like that is not clear. Therefore nobody can accept to be co-responsible for this apparatus. I confirm the conclusion of our Egyptian colleague that the zwikker is no danger for the Lord. The apparatus can, however, be a danger for humanity, but we will come to that later on.'

'I think gentleman and ladies that you forget something,' said the Buddhist, Young Lee Kim. 'If meditation is able to reach God, the zwikker can measure the field of this meditation and in consequence also the contact with God. If the equipment measures nothing, nothing is lost, but if it does, we have to understand what that means.'

'I am not at all frightened,' entered the Hindustani Krishnamurthi the discussion. 'For me God is already involved in everything. Then he cannot find it terrible if he is measured. We are part of his creation and as a part of that we have our own responsibility. The creator himself has laid the foundation for the situation in which we are our now.'

'Mister Chairman, shouldn't we begin with to define God, before we discuss any other implication?' asked Miss Dowson. 'This might be difficult for you since you use the word God or Allah in every second sentence, but it seems to me the nucleus of our discussion. For me the absence of knowledge of God, other than spirit is the cause of so many problems among religions? To help this underway I checked the Koran, the Bible and the catechism of the Catholic Church, but I could not find more than "I am who I am, Exodus III, 14", and "God is revealed to us through his spirit, and no one knows the things of God except the spirit of God" 1 Corinthians 2, 10-11.' Thus we can only talk about spirit, not even image, corps or figure, in spite of the fact that "Let us make man in Our image", Genesis 1, 27. Pope Benedict XVI in his book Bridge to Infinitive specified this as, and I cite: *Nobody can build the bridge by own strength to the infinite. No human being is sufficiently strong to call the infinite his own. No intelligence is enough to certainly devise whoever is God; whether he hears us; how one relates suitably towards him. Therefore, a particular conflict can be determined in the whole religious and intellectual history on the question of God.*[2]

The others looked at her for a long moment of silence.

Archbishop Wiley intervened by mentioning: 'I agree it is a key issue, but let us classify it as follows: the Lord is the axiom of the religions, just as the axioms for mathematics which cannot be deduced from any other assertion or claim, but nevertheless exists. God is accessible through the Prophets, both for the Israelites, the Christians and the Muslims. And what be the creation without creator? Is this satisfying you?'

'No. I admit it is an intelligent reaction, but unfortunately not a clear answer.'

'It's our faith that counts, Miss Dowson, and what's more essential, it might be able to be measured it with the zwikker.'

'Mister Wiley, perhaps my following opinion may help us to approach the subject for which we are here. The largest mystery of our universe is the atom. Immensely small, composed of a nucleus with around itself a sphere with electrons, which is in volume thousand billion times larger than that of the nucleus. The real mystery is that it does not collapse in billions of years since the negative electrons should

---

[2] Benedikt XVI, Joseph Ratzinger, Berührt vom Unsichtbaren, Jahreslesebuch 2005, page 30 December. HERDER, Freiburg-Basel-Wien 2005.

attracted by the positive nucleus. Thus all material in our universe exists of atoms and molecules containing an "active" emptiness which hardly differs from the emptiness of the vacuum between stars and planets.'

'And than the second mystery. If I move my finger, my mind has given instructions to my neurons in my brain to send electrical signals to my muscles of my finger. How is that possible? Somewhere in the emptiness of the molecules of my neurons my mind has the possibility to act. And here we are at what is called the bridge between matter and mind.'

'If so, unless you have another explanation, the Holy Spirit, if existing, has also access to this emptiness in molecules and atoms and to that of the universe.'

'With your permission I will continue with the creation or genesis, marvellously described in the Bible. However, you and I know that this does not match reality, although we may interpret it symbolically. Astrophysicists know that stars are "born" and "die" after billions of years. Their constellations become black holes of compact nuclei with extreme large gravitation force. They may explode again and the cycle of stars and planets starts again or if you like by small Big Bangs. Atoms are again born with an "active" emptiness. From these, life can start after organic molecules have been formed due to external impulses such as sparks.

'With the zwikker we can penetrate in that emptiness in which our mind seems to feel happy. You can understand that we cannot base our discussion solely on what you as religious experts have learned at the catechism and Sunday schools.

'Haven't you ever asked by yourself how people start to believe in God or Allah? Aren't they from their early youth impregnated? Our pre-historic ancestors, who were directly dependent of nature, had reasons to believe in above-earth events and believed in discussions with their deceased parents. But you are anchored in a kind of hierarchical conception, which has become reality by repetition. Wouldn't repetition lead to understanding? An understanding with a feeling of truth! I've used this method with success regularly in teaching natural science.'

'Nevertheless,' she proceeded, 'we cannot neglect this, measurable or not by the zwikker. Too many people have suffered or have been killed in all those centuries, just because they did not understand the phenomenon God or Allah.'

There was a long silence.

The judicious words inspired the others to contribute also to the discussion.

The group of persons exceeded itself, which is rather rare in a meeting, where generally the common intelligence is less than that of the average of the persons present. This phenomenon is explained by psychologists that an individual in a group does not feel himself responsibly for the result. He cannot be addressed on the final result, because a group has a kind of anonymity.

Archbishop Wiley was in the mean time very busy with making notes and he foresaw extra work in the night. As heading he had written: "Zwikker no danger for God, but for people".

In the afternoon the next subject was brought in discussion: "The influence of the zwikker on the faith." Klaus Eckelhof was the next chairman. He was in Germany a prominent Lutheran scientist. Eckelhof carried a small grey beard around his square chin and since his face was almost rectangular, his skull resembled a cube with a beard. He had sharp pleasant eyes, but also a sharp tong, well known with his compatriots.

He introduced the conversation as follows: 'A religion ties up people and the church leaders have the obligation to promote this. A link is created between the persons of a community. This guarantees human relations, although this has also lead to a lot of wars, so-called to defend the proper faith. What I foresee as a conclusion is that the zwikker is a large danger for faith. The apparatus can perhaps show that some people believe only superficially, so that they are unmasked. Everyone of us knows that many believe by habit and education.'

'I myself think positively that I'm a real believer, but that does not mean that I'm critical towards what I see or what I think I see. Since the zwikker may clear this, it means we should attend to that. Perhaps is this enough as an introduction to our second conversation round.' One could see that Eckelhof had more often debated.

The representative of the Lama from Tibet, Jitah Crita stood up. 'You say it quite hard, Mr Eckelhof. If we follow your theory, we must prohibit the zwikker. But we cannot do that. You have read that from the documents. The zwikker can be invented again next year. What can we do then? Nothing! We can only ask our governments to provide guarantees against indoctrination by a little box. But let us not be hypocritical. Don't we indoctrinate day after day children and adults? How would be a complete blank creature if he was not influenced from outside? Perhaps would he be closer to God than we? You know some bible texts which talk about that, isn't it? We, as religious representatives, usually do not wonder on the impact of indoctrination by ourselves. Sometimes we are even happy with it, or think we are. Is "de"doctrination of our indoctrinated values with the zwikker a danger? Against indoctrination with the zwikker we must act, however, but is that really new? A question to be discussed too.'

'I do not believe, that we must put our faith in discussion,' spoke Jami. 'More than any other religion, the Islam has survived since it is based on the Koran. We cannot deny that sects in religions have caused many separations. But this frequently occurred, because someone wanted more power.'

Eckelhof seized the control of the discussion. 'Let us examine at this moment only how we must formulate our conclusion concerning the faith and the zwikker. I ask archbishop Wiley if he has a proposa. I saw him writing energetically.'

This was accepted and the texts made by the archbishop inspired everyone to find a consensus.

The next three days the participants debated on subjects such as: Is it possible to measure the subconsciousness of dying people? Is the soul the basis of the subconsciousness? Can one allow the use of the zwikker for testing speech and meditation? Should we perform tests on relics, sacraments, holy places and praying the rosary?

Although each participant was curious to know the result of such measurement, they feared that the religious notion would be disturbed unnecessarily. There were always people who regarded something like that as the work of the devil. And history had learned that mass psychosis and fundamentalism was born quicker than being oppressed.

After three days, the meeting had come to the subject of "the use of the zwikker as an indoctrination instrument". Nobody would, however, assume responsibility to launch this as a positive conception. They feared being stigmatised by the home front when they recommended the zwikker as a useful instrument

Thus, although the group had started with large enthusiasm and certain solidarity, the meeting ended as a night candle. The end document was phrased in very careful terms. After the enumeration of the discussed points and the difficulties which one saw in the use and application of the zwikker, it was concluded that governments should limit the use of the zwikker and that the religions should remain mutually in contact. To promote this they proposed to set up an international meeting office. The office should be close to the UN building.

New York

Dolores Guerrero smiled. The report of the meeting in Alberta lay in front of her. Here again a new commission had proposed. That almost always happened. The group had hardly pronounced a verdict. She knew the commission syndrome and in her organisation more commissions were established than erased. The only manner to keep the number of commissions in hand was to limit or stop funding.

'I may be able to help them,' she thought. 'They can use our secretariat. That is already eighty per cent of the costs. I will inform the Archbishop of Canterbury.'

She found the treatment of the subjects interesting. Anyway the meeting in Alberta had not rejected the zwikker and she could work further with them, if necessary. The information about the universe and the emptiness in atoms by Miss Dowson seems to be very interesting, but too premature to make it public. That would only disturb the discussion.

From Russia only one report had come in. The situation in Uzbekistan remained stable. No zwikkers had been found and none of the compensated, including Raskadov, showed any consciousness convalescence.

The first monthly report of the medical group in New York was interesting. Anyway the discoveries seemed to be innovative. No doubt, a request to make indoctrination experiments would follow. She better should be prepared for the authorisation.

Washington

The feelings of Peasley concerning these reports were different. In front of his desk sat Brill Dowson the Humanist, who had participated in the Alberta meeting.

'You have talked a lot, sister, citing bible texts. Have the others converted you?'

'It seemed wise to make first these remarks, Sir. If the discussions concern consciousness related to God or their religion, they have problems, not me. I haven't any desire to help them thereby. When the zwikker becomes the topic, I will join the discussion automatically.

'I'm not so certain about that, be aware of that! I thought there has been only one Big Bang for which a Nobel Prize was awarded. And you suggest in your own something different? Are you also explaining that the increased extension of the universe is not due to the dark energy?'

'Indeed Sir. That extension of the universe is due to attraction by older star configurations which have been formed by older Big Bangs and died in black holes. It has no sense to believe in pushing dark energy.'

'So except of playing a kind of Jung, who claimed a common conscienceless, you play for Einstein.'

'I'm indeed very proud of my conclusions about the genesis of our world with more than one Big Bang, the living emptiness of atoms and only attracting gravitation.'

'Young lady,' barked Peasley. 'You know very well that this is none of your business!' 'You are here in the first place to avoid that these illuminates will hamper the US policy. We must also prevent that somewhere insurrections come against the zwikker or that the thing is banished, as far as you can speak of spell or taboo. You aren't for nothing on my payroll.'

'Don't threaten me, Sir. I don't let myself impose an opinion, neither by anybody nor by you. Of course I keep in mind the US interests. They have not yet deprived me of my patriotism!'

Peasley tried to control his anger. He thought a moment to kick her out, but since she was already a member of the Alberta-group, he controlled himself. 'I m not used to such answers, young lady, take that well into account. The next time you have a larger input and keep in mind that they aren't going in the wrong direction. You can go.'

It lasted some time before he could concentrate on the two other reports which lay on his desk. 'A pity I've no fellow in the Moscow centre and in the New York hospital. I don't like it at all.'

9

Is courage required in science?

New York

The scientists in the Central Hospital had indeed got an authorisation to carry out step by step indoctrination tests at the condition that each test should be reversible. In other words one should be able restore the old subconsciousness.

The laboratory had for this purpose obtained from Moscow a new zwikker sensor which could be focused at components of the brain. Professor Larson, the leader of the group, had proposed to be the first volunteer. From his earlier made total spectrum they had already found a part which deviated from that of the others and which possibly could be related to his speech defect. They compared this with that of a compatriot of Larson who had been especially commissioned for this test. Alfred Petterson, as he was called, was a psychotherapist and strongly interested in the zwikker research.

They assumed that the speech consciousness of the two Scandinavians would be approximately identical and the stuttering made the difference. They hoped it was a matter of consciousness and not of a nerve deficiency between the brain cells.

The tension was on all faces when Marga Stalma, the stammer expert and the Boris Klemarov the technician established themselves behind the zwikker. Professor Larson had gone in the experiment room and lay down on the white table. The new sensor was assembled above his head and could be moved by Marga Stalma and Klemarov.

Marga Stalma focused the sensor at the part of the brain which contains the speech centre and started the zwikker. All looked concentrated at the screen. At first nothing happened.

The screen remained virtually empty, except that from left to right a vibrating line moved. Klemarov increased the sensitivity of the zwikker, but this resulted only by an increased vibration.

'Is that sensor O.K.?' asked Marga.

'According to me the sensor must be all right, because he worked very well during a test at some distance of an object in Moscow. I don't understand.'

'Would it be possible that we measure in the wrong way?' asked Jean-Pierre Petit. 'Perhaps the equipment is not able to measure such details as stuttering in a speech centre?'

A discussion started to which also Professor Larson joined. 'Perhaps we must approach it co-co-completely different,' he said.

'Of course,' exclaimed Marga. 'We go much too rapid. We must approach the cause of stuttering in same the manner as we do by means of psychoanalysis, without a zwikker. One of the methods we use is hypnosis, and what can be done under hypnosis might also be possible with the zwikker. Only the patient does not need to be now hypnotised. If the cause lies somewhere in subconsciousness, we must search there. The zwikker can't execute brain research. We must restrict us to the consciousness research.'

'This implies,' interrupted the psychologist Bertram Slinhald, 'that we must return to the spectrum which we already have of Professor Larson and compare that with the results of psychoanalysis which we have made afterwards.'

'But there are thousand of possibilities!' exclaimed Petterson, who feared, he had come for nothing.

'I don't believe that,' said Marga, 'the stuttering of Professor Larson has started at a certain moment in his life and generally this happens through stress to which the person has been exposed. Is that true Professor Larson?'

This one confirmed that. 'Do you want to hear my tale whi-whi-which I told frequently to the therapists who t-t-tried to cure me?'

Larson had grown up with two older brothers and three sisters and their largest pleasure was to imitate people. Since the gatekeeper of their neighbours was a stutterer, they had frequently imitated the man. To the pleasure of his brothers and sisters he had become the best stutterer. When once he had been laughed at by school children, his stutter became permanent. Only if he sang he had no problems.

'Now we know where we must search approximately. In the subconsciousness of this period, which is in the left part of the spectrum.'

The others listened curiously. Perhaps Marga was right. Klemarov looked up the file and brought it next on the screen. He let passing the groups of vertical lines from left to right, until Marga called suddenly: 'Stop! Hold the picture, Boris. And divide the screen in two parts, one above and one below and have that of Professor Larson in the low part. Do now the same for Petterson's spectrum in the above part.'

Klemarov did what she asked him.

'Move that of Petterson slowly to the right. Not to fast, please. Yes, this starts to give something that looks identical.'

'Stop, Boris, this is it. Still a little bit backwards. Yes, splendid!'

What they saw were two almost identical groups of vertical peaks which differed at the most in length. There was however one group which showed larger differences. Marga indicated at it.

'Professor Larson, do you agree that we compensate this part of your subconsciousness with that of Petterson?'

'No pro-pro-problem, go ahead. I trust that it will suc-suc-succeed.' Professor Larson went to the other room and lie down on the table. Klemarov replaced the new sensor for the original and selected with a "block-system" the component of both spectra. On the screen the two components were placed between two vertical dotted lines. Klemarov gave the zwikker the task to calculate the compensation spectra of Professor

Larson. He had now made three horizontal sections on the screen, at the top that of Petterson, in the middle the one of Professor Larson and below with the compensation pattern as a mirror picture.

'Is everyone ready?' he asked

'Wha-what do you mean with everyone?' shouted Professor Larson from the other room. 'You are only treating me, do you?' He appeared slightly nervous and had small perspiration droplets on his face.

'I will put now the zwikker on "compensation" and according to my calculation he will pass through the block of the spectrum in exactly five seconds. As a matter of fact he stops automatically when he is ready. Klemarov pressed a button and a red light illuminated on the sensor above Professor Larson's head. After exactly five seconds it went out.

'Is that all?' asked one of the bystanders.

'No, be patient. The zwikker is now counting. You will see it in a second on the screen. Or better you see it disappear.'

The computer gave a small peep noise to announce that he was ready and slowly from left to right the block disappeared in the middle section. Nothing remained at the place between the two dotted lines.

'Is everything O.K., Professor?' called Marga slightly worried. They saw Larson opening his mouth but there was no sound.

'Oh, heaven!' called Bertram Slinhald. 'He is beyond words now!'

'No panic, please!' shouted Klemarov above the arisen tumult. 'Pay rather attention to what will happen next. I will now start with the procedure to "in"doctrinate the spectrum-"block" of Petterson to Professor Larson. Let us wait first for the result!' Professor Larson, please continue to lie quiet, it takes only a few minutes.'

Professor Larson put his thump up and closed his eyes. He had really panicked. He had seen for his eyes all kind of pictures from former days disappearing as in a dream in the distance, as the rails from a railway to a point. It finished with small thud of which he was badly started. When he wanted to say something his throat was blocked. As one large and ongoing stutter. The sweat had broken out and he had desperately looked at his younger colleagues for help. The words of Klemarov had him somewhat reassured. Perhaps everything would turn out well.

Klemarov gave the zwikker the commando to transmit the spectrum-block of Petterson to Professor Larson. On the screen in the upper section a red vertical line appeared which moved slowly from left to right. Each line of the block became a moment green. The whole procedure lasted fifty seconds.

There was a large tension in the room. One smelled perspiration and some scientists held their breath. Professor Larson continued to lie with his eyes closed as if he was under narcosis.

'How is it with you, Professor,' whispered Marga. 'The experiment is completed, therefore you can open your eyes and rise from the table.'

Professor Larson opened slowly his eyes. Set up straight and rubbed with his hands in his eyes. 'Where am I? And who are you all there behind that window?'

The complete group said no word and looked at Professor Larson as he was a peculiar fish in an aquarium. The fish looked back just as interested.

'Hey, Klemarov, what I'm doing here and can I leave the room?'

'He talks normally,' called Marga, 'and with Petterson's accent! Boys hip, hip, hurrah for our Professor and for Boris. Hip, hip, hurrah...!'

The others awoke from their apathy and joined with even harder: 'Hip, hip, hurrah...!'

'Why is all this upheaval?' asked Professor Larson when he came from his transpiration room. 'Why you shout this way with joy?'

'You don't stutter any longer, you hear that, you can speak normally!' called Marga.

'What stutters, who stutters?' asked Professor Larson surprised.

'You did that yourself, you really don t remember?' called Bertram Slinhald.

'I am confused about what you are talking about. What I do remember is that I saw all kind of strange people whom I had never met before. I was a small lad and there was a large sister. She was called Johanna.'

'That's my sister,' called Petterson. 'I have an older sister who is called Johanna. Can you describe her?'

'Blond with golden spectacles; she carried a clamp to rearrange her teeth and she called me Allie.'

'That's indeed my sister. She called me Allie to tease me. In your consciousness you have gained a sister,' said Petterson.

'You must explain it later to me in more detail,' spoke Professor Larson. 'The last thing I remember is that I should undergo a test with the zwikker.'

'We will show you the video pictures which we have made of the test, so you can see it for yourself,' said Klemarov and started the threetel-video from the moment Professor Larson entered for the first time the experiment room.

'Interesting,' he declared. 'I don't remember this happened. I really stuttered, and now no more. A part of my memory must be erased. That should not have happened. Perhaps it returns. Memory and consciousness are two different things, but probably they do have links. Youngsters, congratulated with the result. And thanks. I no longer stutter, although you say that my accent has changed. It seems to me useful that we must analyse this test carefully before we go further with other experiments. The fact this one was successful, is not a guarantee for the next one.' Professor Larson had recovered and took again the leadership.

'In fact only a hundred percent proof can be given when we can bring you back again in the old situation,' said Marga, 'that one with the stutter deficiency. Wasn't that the condition to carry out our experiments?'

'No, no, I'm persuaded of the functioning of the zwikker and I am not in for such an affirmative test! I'm too glad of the result and one never knows what kind of secondary effects such a control test can have.'

'You nevertheless don't mean the deviations which I have? Although, if I have them, then you are just as normal as I am,' said Petterson.

Everyone burst into laughter and the tension of the last hour slimmed down. There arose a lively discussion between the pro- and contras for an extra control, and even to induce the stutter to Slinhald. Since the zwikker had saved the erased spectrum of Professor Larson this would be possible. But Slinhald was very much opposed and since there were no other volunteers, the general poll was to accept that the experiment was a success. The scientists continued for several hours their work on detailed interpretation of their spectra and the psychoanalyses. Almost three hundred peaks in the total spectra had been identified.

They had the feeling that they could now pass on with confidence to the second experiment which concerned the study of the subconsciousness of a Mongoloid child. The authorisation for such an experiment had already been obtained. A boy of ten years was selected of which the parents hoped that his intelligence could be improved with the zwikker.

The Mongoloid boy Jessie Everton had a two year's younger brother called Peter, who was completely normal. According to Professor Larson who had examined them, there should be some conformity in their subconsciousness. Mr and Mrs Everton were end forty, which implied that they had got their two sons rather late. It had been a great sorrow that Jessie was a Mongoloid. They had not asked for preliminary analysis, out of fear for possible errors in the diagnosis. They wanted afterwards no more children, but Mrs Everton had become pregnant again. To their large joy Peter was a healthy boy full of life, desire and initiative. He had compensated much of their sorrow. Although Jessie requested by his handicap much care, they were a happy family.

The next Wednesday the family Everton stood already waiting in the hall. Professor Larson welcomed them personally. They followed him through the long corridors, with Jessie full of joy. He spoke poorly and stammering, but his face radiated. He liked white walls and he always wanted an adventure. His parents had told him that they would play a nice game with people in white coats. He himself would have also such a coat.

In the laboratory Jessie shook hands with everyone, something he enjoyed to do. Peter remained close to his mother. He knew better what it was all about. His mother had explained him that they could perhaps cure Jessie. She had quoted a famous fairy tail that Peter had frequently read and Peter bravely wanted to help.

To make the experiment more attractive, Marga Stalma had decorated the detection room and each child would receive a nice present at the end. First Jessie was allowed to enter in the detection room. He was dressed in the white coat and as instructed, he lied down obediently on the table. Klemarov had already warmed up the zwikker and pressed the button to start the automatic scanning. He had told Jessie that he should do as if he slept and fortunately Jessie behaved perfectly. At the moment when he started to make some impatient movements, the zwikker clicked off. Marga took Jessie from the detection room and all came looking at the spectrum which appeared slowly on the screen.

The spectrum deviated clearly from those which they had measured with themselves. Nobody dared to make a remark. They let act the pictures slowly by themselves. Only Klemarov explained to Mr Everton what the peaks and lines meant.

'And now you, Peter,' spoke Professor Larson. 'Do just the same as Jessie. Don't move when you are on that table and close your eyes. When you hear a click it is ready.'

'Must I, Ma?' asked Peter looking at his mother.

'You don't have to my boy, but we want to help Jessie and perhaps these people can do that. You feel nothing, you must think of nothing, can you? The sooner we are ready you will have your nice present.'

'And where is my-my pre-present?' stammered Jessie. 'I've been go- good.'

Ma Everton smiled. 'Here my darling. Look in this bag.'

Jessie tore the bag open and saw a brown bear looking at him. He unpacked it and detained it stiffly against himself.

'Will I also get something like that?' asked Peter.

'Sure, son, this way,' said his father. 'I will bring you to the little room? Here is also a white coat for you.'

This persuaded Peter and a moment later the detection could start. Klemarov pressed the button and after a something longer time than that of Jessie the zwikker clicked off.

The click did leap Peter from the table and he returned quickly to his parents. Pa Everton handed him his gift and to Peter's joy it was a plane which he had wanted badly.

The others looked at the screen and what they saw was a pattern that looked quite different from that of Jessie. It resembled much more the spectrum of adult people, only the hillocks were less developed.

Professor Larson spoke first: 'It seems to me that the hypothesis of Pierre is confirmed. Jessie has a subconsciousness field smaller than that of a normal child such as Peter. There are nevertheless some agreements because they have been raised in the same family. Up to that point your hypothesis is correct. But what do you propose next, Pierre?'

'We have here two spectra,' started Pierre hesitatingly, 'one with a damaged pattern and one with a healthy pattern. We must explain to the parents that it is possible to introduce certain patterns of Peter's consciousness field to Jessie. I propose that we first take a small part of Peter's spectrum. The test lasts then no longer than three seconds.'

'Is there any danger?' asked Ms Everton.

'We don't know for certain Mom, but we can show you the video pictures of similar test we have done with ourselves. The healing of stuttering for example.'

'Must that?' said Professor Larson, who preferred to keep this secret.

'If you say that the risk is absent, Mister, then you must do the test,' concluded Pa Everton. 'Finally we came for this. It's not necessary to show the pictures of the other tests. I might not understand them and it may lead to misunderstanding.'

His attitude pleased Professor Larson. He gave the signal for the second test.

'Must Peter go again in that room,' asked Ms Everton.

'No, no, I've stored his spectrum in the computer and I can simply use that.' It was Klemarov who answered her.

Marga brought Jessie to the little room and Klemarov started the zwikker. He took that part of Peter's spectrum where Jessie showed virtually no patterns. The click followed thirty seconds after he had pressed the button, and Marga called again Jessie who had kept himself very quite.

Klemarov measured subsequently again the spectrum of Jessie and to everybody's surprise and joy, a pattern appeared similar to that of Peter.

Marga asked Jessie to come over, what the boy did delightedly. He found it obviously a nice game since he got all attention of the people around him.

Marga questioned Jessie whether he had dreamed something special.

'Oh yes,' spoke Jessie, 'we had summer holidays and we built on the beach some sand castles. And Pa found I did it terribly. I got a flag which I could put on the castle.' He looked around, slightly astonished about himself. He had spoken differently, no longer stammering.

A sob sounded in the room. Ma Everton had started crying. With tears in her voice she said: 'Sob, it was Peter who had done that.'

'No Ma, I saw it myself,' repeated Jessie. He looked around whether there was someone who would praise him. 'I was it, say Pa?'

'Of course my boy, it was you, and you did it very well.'

The group started to discuss this result. It was proven that one could introduce a better subconsciousness in a mongoloid linked to a connected memory. They left for coffee and tea to celebrate this and beat father Everton on his shoulders.

He, however, was not yet persuaded and said: 'Misters, my wife and I have had already many deceptions. Don't take it evil when I don't yet believe in what you call a first result. We know that mongoloids have a genetic deviation and that this deviation cannot be healed. What you have implanted just now in Jessie cannot change that according to me.'

A deep silence fell. In these words resounded the worries of ten years parenthood, whereas the scientists had only thought of reaching results with the zwikker. A feeling of respect rose in them and Professor Larson was the first who spoke.

'We don't take it evil, Mr Everton. We can understand your feelings, but from the moment we worked here, we hoped so much of doing something good. Your son is really no object for us, please. We respect him as a loved human being. It was only as a result of our tension that we reacted joyfully.'

'Does someone of you think that Jessie could become a normal child by giving him the subconsciousness of his brother?' asked Ms Everton.

'I don't think for hundred per cent, but very possible to a certain level,' answered Pierre. 'For this reason we must check soon if Jessie still has that part of Peter's spectrum. If Jessie's brain can retain these consciousness fields to some degree, a repetition of the treatment may force results, even if each time there is a drawback. Let us recheck in half an hour.'

'Perhaps you must do that,' said father Everton. 'And if there is nothing more, we go home.'

All members of the groups looked earnestly. They couldn't blame Mr Everton that he would not let his son be used as a test rabbit. The Everton's had cherished already too frequently false hope.

Marga asked Jessie whether he would again go in the little room and tell them what he had dreamed. Jessie, to who all commotion escaped, nodded and went smiling to the table. Klemarov started the zwikker and repeated the usual test procedure.

When Jessie returned by himself, since nobody had called him, he found them all staring at the large screen. The right hand pattern of Peter had been slightly reduced. It was also noteworthy that Jessie stammered again slightly.

Mr Everton stood up. He had seen enough and spoke: 'Come Ma, Peter and Jessie, we go home.'

'No- no- now already?' said Jessie, 'You-you said I could play here?'

Mrs Everton hesitated. She had seen the temporary impact of the zwikker on Jessie. She had had so much hope. That boy deserved more. In spite of his weak gifts, he was happy and liked to tease people in a gentle way. Perhaps the test had been too short? Was that perhaps the cause?

These ideas passed through her head and Marga, who was paying attention, understood what passed through her.

She went to her and said: 'I had for eight years a nice man and a little son. They were both taken away from me through an accident. I can sympathise with you. The loss of hope is very terrible.'

She caught her hands and continued: 'If you want, we can still do one test. We transmit the complete consciousness field of Peter to Jessie. We have seen that only little disappeared, so there is also a chance that the rest might remain. What do you think about that?'

Jessie's mother looked at her husband. 'Well pa? It comes more because we bother ourselves more than does Jessie. He finds this a nice game. Perhaps he feels something that he has never experienced before?'

The face of Mr Everton got a gentle expression. He loved his wife.

'Well then,' he said.

Jessie who had tried to follow the discussion, jumped up. 'I ca- can play further, Pa? Hurray, hip, hip!'

Klemarov used the occasion to load the zwikker with the full spectrum of Peter and whispered then to Marga: 'Bring that boy rapidly in the little room and ask him to lay down completely quiet for as long as we ask. Tel him that he soon can select something beautiful for his bear in the child shop of the hospital.'

Marga did what Boris asked. Jessie did not move a finger during the complete test. Boris had put the zwikker on the slowest speed and it lasted five minutes before the spectrum of Peter had been transmitted.

At the click Jessie leaped up and came outside. 'Will we go to play together scrabble, Peter? And can I select also such a beautiful plane has you have?' The words which he said had been almost pronounced with Peter's voice.

Before the emotions would possibly play a role with the parents, Professor Larson called upon them: 'Mr and Mrs Petterson, please come along with Jessie and Peter to the child shop and we won't disturb you any longer. Your nerves have been already tried too much. However, I would appreciate if you could return with Jessie in a month's time. I have now full hope that we can improve Jessie on the long term by regularly treatment. It has often happened that unused parts of our brain can take over defected ones. And who knows this will happen with Jessie. It will take time, but the experiment has proven that it is possible.' He requested Marga to accompany them to the child shop to buy the promised gift for Jessie.

A sigh of relief went through the laboratory. Professor Larson ordered coffee and cookies.

10

Is patriotism authentic or phoney?

Washington

Peasley was impatient. He expected Spone back from Uzbekistan, but he hadn't shown up yet. He hated when people came too late. He regarded the report of Spone on Uzbekistan, but the contents were good for nothing.

Someone knocked at the door and he called: 'Come in!'

The head of Spone appeared. 'Do you expect me now, or shall I come later?'

'Don't dawdle man, I'm waiting already ten minutes for you. Why you made such a worthless report. Do you think I believe the content? Everyone in that country wears softy cheerful smiles and they are kind to each other. Did they show you around with a sect?'

'Certainly not, Sir! I was free to go wherever I wanted. Everyone was helpful. Do you know what I think?'

'Cut it, I want only the facts.'

'But these you find in my report,' reacted Spone offended, 'the population in Uzbekistan is really as I have described. Relaxed, helpful, sociable, in short, without complexes. Delightfully I felt, I was there almost on holidays. Only those weak heavenly smiles, I couldn't get used to it.'

'So, holidays, and you didn't get used to smiles. But there must have been people who were different?'

'Only the Russians who came visiting. Very crazy situations where Muslims kneeled together with Christians in Orthodox Churches and received hosts. They mixed all up as if it was a child game. Yes..., that's what it was. The people had something childish over them. However, they spoke normal to me in a bar. The Muslims drank quietly vodka, although they couldn't stand it.'

'What did those Russians further?'

'Looking and writing notes, just as I did. There were no tensions, only too many drunken people.'

'Did they run around with zwikkers to survey people?'

'What do you mean, zwikkers? That's forbidden, isn't it?'

'Mug, you are in my service long enough to know that something like "forbidden" does not exist?'

'I haven't seen anything of that kind, nowhere where I came were Russians with recording equipment which could be hold for zwikkers.'

'Seems me a bit thick. Well, I know enough for the moment. You stay in your office and you collect all bulletins concerning that country in the press and Internet. As soon as there is something particular, let me know. I want particularly be informed on how long that hypocrisy persists. Scram it and do your work.'

Spone disappeared with a painful grin on his face.

Peasley hoisted himself from his chair and paced up and down in his office. 'Cheerful smiles! My foot. There is something behind it. Religious people behave also this way. Certainly those who think they are blessed. But I know better. When they turn around that holy smile has disappeared. It is urgent that our zwikker becomes available.'

He did not love uncertainties. It was already a month after he had heard from Travenkov that the zwikker plans were also in possession of the West European Union. The people he had put on De Beaufort and Holthus had reported nothing but normal. Holthus was as usual at work. He made no suspected calls and received on his bank account only his normal salary. Peasley did not understand. That Holthus had acted on idealism, he did not believe. And that De Beaufort would do nothing with the plans, he could understand even less.

Each moment he expected a call from President Smith to this account, but nothing had happened. Smith must have read about the arrest of Spone. But even about that the President had not approached him. What was going on?

He suspected, however, that President Smith did not trust him entirely. But someone having his own responsibility you don't carpet for each press report. Smith's conception was that Peasley had to execute his task in his own manner and if Peasley thought that somewhere a security system was not good, it was his duty to check this, especially after that robbery of the zwikker by Uzbekistan.

Peasley tapped on his desk. The head of the special laboratory should have come at 10 A.M. and it was five minutes after. He wanted to know how far they were exactly with their research and if they had assembled already a complete zwikker. Time was pressing. Each moment counted. Any moment he could receive an order to stop his research.

He had read the first reports concerning medical therapeutic experiments in New York and concerning the technological developments in Moscow. He found the progress but slack. Especially that of the medical research. What is the benefit of a revolutionary invention if it is only used for helping mentally disturbed people? That thing can influence complete populations. That's always the case when commissions and politicians make the decisions.

The intercom buzzed. 'Yes!'

'Dr. Sheiley and a colleague for you, Sir.'

A man with a goatee was let in. 'Take care that I'm not disturbed, by anybody.'

He didn't stand up when the head of his special laboratory, Ron Sheily, entered.

'How far are you? You know that we have to hurry and spent no time on the finesses of the apparatus. Only that it works. And?'

'We continued working the whole night, Sir, to show you the first result. The reason I'm somewhat late. With your permission I will call my employee who can show you the first US-zwikker.

Peasley muttered affirmative and Sheily showed in a man who carried a small suitcase. This one greeted his boss politely, but Peasley only signalled to put the suitcase on his desk.

'Do you need him?'

'Not in the first place, Sir.' Upon which Peasley requested the employee to leave them alone.

'The lesser, the better. Why it took you almost two months to assemble the apparatus? Hadn't we the complete design plans? Something like that could be done in two weeks, I would think?

Sheily frowned piqued. 'To the plans was lacking a small, but essential information. First the plans were in Russian and had to be translated. Secondly the plan for making the gyroscope referred to an article in a Russian technical magazine. Although this was obviously public information, we could not find the magazine anywhere in the US. A week has passed before we found it in a library in Berlin. We could hardly request Moscow to send a copy.'

'Well, well, how far are you now?'

'This is our first zwikker. And one which is better than the original. The plans of Moscow assumed still some old-fashioned electronics. We have applied the latest novelties.' Sheily looked as if he expected a compliment.

'Does it works also better?'

'Not only that, but also farther, both for consciousness measuring and on gravitation.'

'What do you mean with farther, farther away, or by means of radio or TV, or to the moon?' Peasley became always impatient when someone gave a step by step explanation.

'Sheily, come to the core and do not take me for an imbecile who must be explained slowly.'

'The zwikker can measure a consciousness field or the field of gravitation up to two kilometres distance. One must only have a sensor which can zoom in between near and far. Transmitting is possible to the same distances, and for consciousness also by means of the threetel. The quality of the spectrum is somewhat less at a large distance, eighty percent approximately. Of course one can only measure and compensate consciousness by means of direct threetel emission. For gravitation this is not possible.'

'The first is nothing news, I knew that already with Uzbekistan. But why he is it not possible for gravitation?'

'The threetel cameras cannot measure a gravitation field.'

'I don't believe that, it's not a matter of measuring but passing on of fields. I have read everything what has been written about the zwikker. According to me one can influence both by means of the TV. I know I'm right. You must try that of the gravitation another time. How far are you with a detection apparatus and an anti-zwikker? We must have these two apparatus within some weeks.'

'That rapid already?'

'Yes, as rapid and still more rapidly if possible. I cannot you tell why, but it is of State importance.' He looked just aside to the American flag as if to confirm his tale. 'How far are you?'

'The detection apparatus is in advanced state. We have found that the zwikker is transmitting a field with a large range. So far we had no particular equipment which could detect this field, but one of our technicians came on the idea to use an existing plasma apparatus for metal detection for this purpose. He had noted that this plasma apparatus gave derogatory results at the analysis of metals in air or water when the zwikker was working nearby. The plasma apparatus could be a zwikker sensor. At present we can discover the zwikker at ten kilometres distance if it is switched on and detects. The distance is hundred kilometres when the zwikker "emits". Our problem is only how to reduce the size of this apparatus so it can easily be transported. For producing the plasma phase, a very high temperature is necessary, and this requires much electrical energy.'

'And when is that solved?'

'In the coming months, not earlier. And that counts also for an anti-zwikker. We think to use the principle of the zwikker itself. Since the zwikker gyroscope is sensitive to the field between spirit and matter and to the field between matter and matter, we must build something which disturbs this sensitivity. In other words....'

'Yes I get it, with a compensation field of some microseconds you disturb the zwikker. And if you have a malignant apparatus, you disorganise the zwikker by destroying the gyroscope. I have that guessed all right?'

'Approximately.' Sheily could not be put off. He knew that his boss wanted to show the cleverest.

'And when is that thing ready?'

'You must still have some patience. We work day and night. With the current restricted staff we cannot do more. People also have to rest now and then.'

'Why then not have more people at work? Don't use such a rotten argument with me. You can have a vast budget. If necessary, you can use the zwikker to indoctrinate new people as that is called.'

Sheily reacted shocked. 'We still don't know what exactly the influence on people is. I read that the group in the Central Hospital, after an initial fast start, had some kind of problems!'

'Then I hope that someone else on the street will not indoctrinate us and let us dance as monkeys. Sheily, your pretexts may have large consequences. Do you want that our American life style becomes zwikkered by someone who can handle such a thing? And the beautiful fact is that we will like it! Think about that once more!'

He was silent for a moment. 'You arrange an extra group of people entirely on the application possibilities of the zwikker. These fellows must work faster than those in the New York hospital. I will send you today my most skilled people. You lock them up for a month with the zwikker. Let them practise on prisoners, gangsters and soldiers. You have exactly one month's time. Longer is not possible for several reasons which you can imagine by yourself. You are an old hand at the service. Money can be no restriction. You can get what you want, although we have to justify it later on. Then the US has at least everything concerning the zwikker in house and we are no longer dependent from a couple of softies.'

Sheily's goatee started to move up and down. This was always a bad sign for people who knew him. He got the notion that something was fishy. What in fact had Peasley in mind with the zwikker and his patriotism? He had not at all spoken about the official policy of the US and whether the President had asked him to solve the zwikker problem. With a personal zwikker he could become a dangerous person. Sheily said no more.

Peasley interpreted this as an affirmative, pulled himself from his chair and walked around the desk to the zwikker. 'How many of these things you have already?'

'This one is almost completed. It is lacking still some essential components. But to answer your question, we can provide in two months four complete zwikkers.'

'Take this thing along, and deliver me as rapid as possibly a complete one and faster than in one month, not two. I will show that to President Smith. The other ones you keep for your research. Under no circumstances any zwikker may leave the building without my agreement. Am I clear?'

'More than clear, Sir.' He stood up, placed the zwikker in the suitcase and left the office.

Ron Sheily had not told the truth to Peasley. The zwikker, which was on his desk, was already operational. But the way the conversation had gone, had hampered him to give a demonstration. He hasted to the laboratory and asked the long employee, who Peasley had sent away, to store away the apparatus.

The long employee, called Ben Corward, returned to Sheily and asked: 'How was the big boss? You look very pale. Has he bellowed again?'

'Indeed, but what he said is more serious. He wants to know how far we are and I reported that our progress was less than in reality. He now thinks that we can provide in two months four zwikkers and that the detection apparatus and anti-zwikkers are still in development. Whereas, we have all three already available. What still is lacking is a detection apparatus that can be used from a plane. It's my consciousness, Ben, that worries me. Peasley does not seem to have something like that. I have the impression he is busy for himself. What do you think?'

Ben, who folded himself up on a chair, looked taken aback. He was still too short in the service to permit criticism on the big boss. The statement of his direct chief and friend assaulted him entirely.

'But that can't be true? Peasley has to stick to the law and to the authority of the President. You're certain?

'Not really, but if I put everything on its place, my hypothesis is correct. Listen. By the end of September came the news that the zwikker had been invented. Due to pressure by a number of countries, and our President was in agreement, it was decided that the zwikker would only be used in two places, Moscow and New York. An international supervision was established and only Interpol occupies itself with safety aspects. Then there is a bulletin that a US spy is caught in Moscow, called Spone. Spone gave as excuse that he had to test the security system of zwikker research centre outside Moscow. In the weeks that follow there is no further news. We only receive reports from the Central Hospital of New York and from Moscow. But, and now it comes, just after Spone is released in Russia, Peasley calls me and shows me the construction plans of the zwikker which were in Russian language. He mentioned a number of patriotic arguments that the US, as the strongest country in the world, cannot permit itself to be bullied by criminals who can get hold of a zwikker. We must build therefore such a thing by ourselves and moreover a detection apparatus and an anti-zwikker. And..., not one word from the President. If Peasley wants to use the zwikker to conquer power, where do we remain? And what is more, what must I do?'

'Why must you do something? Do what you're told, then you won't get problems.'

'That's what I have told myself several times, but it does not help. Perhaps I must use the zwikker on myself, so that I can live more quietly with my consciousness.'

'And if you ask to speak with the President?'

'That's another difficulty. He is protected by all possible hierarchical structures and I can never pass those without Peasley knowing it. I really don't know what I should do. I would pay gold when someone could give me a good suggestion.' Sheily looked at him for a relieving word, but he only saw a long question mark folded in a chair.

'Perhaps I have to become the first criminal who will use the zwikker for personal use. I only hope it will not release a spirit from the bottle. I see for the moment no other possibility. I feel I must do it by Christmas. Ben, I think you must either support me or stop me. There is no other choice. It is not yet so urgent, but do think about it and give me your opinion. I will respect your decision.'

Moscow

In the zwikker centre in Moscow Yuri Kaspadov had a discussion with President Bolotnikov. The head of the Russian Security Service, Nicolai Yesin, accompanied him. Kaspadov had first showed him around in the laboratory. He had presented them to the international group, called the IZRG (International Zwikker Research Group), which was since September very active. Then he transferred Yesin to the head of the security section.

Remaining alone with Professor Bolotnikov, he gave an overview of the achievements.

'To answer your first question, Sir, our satellite detection system has become operational last month. With "our" I do not mean the international group, but the Russian one. The IZRG is not yet informed. We hope, however, that the international group will have an anti-zwikker ready as well. They work on a detection system for short distance.'

'Is the detection system of the IZRG much different from ours?'

'Not much, they have a derogatory detection head, but the principle is the same. We have made slight improvements to have the satellite sensor operational.'

'And do you have already some results?'

'Yes, Professor, they are given on this world map. The satellite rotates in circles around the earth and can cover the whole sphere. Since the rotation of the satellite is somewhat more than an hour, we cover the whole globe in approximately sixteen days.'

Bolotnikov examined the map carefully. A number of flags had been placed on the card which designated Moscow and New York. There was, however, one flag that attracted his attention.

'What does that flag mean? There aren't any other zwikkers, do they? Isn't that Washington, or am I wrong?'

'We try to check it. It concerns a single measurement. We have checked whether it wasn't caused by an error in our detection system or by cosmic rays or radio waves. So far we haven't been able to observe any abnormalities in our satellite communication systems. We must therefore assume that the measurement in Washington was correct and that somewhere a zwikker was working there.

'Could this be related to the robbery of the plans which you have reported in October?'

'Perhaps.., perhaps. We can ask Yesin which information he has accumulated and study together our combined information. This was the main reason why I have asked you to pay us a short visit. Fortunately you have accepted the invitation.'

'So what we were afraid of has happened. There is somewhere in the world a "wild" zwikker and the question is how much damage it has caused already?' He requested Kaspadov to call Yesin with his pocket card.

'Yes,' replied Yesin.

'With Kaspadov, Mr Yesin, would you be so kind to join Professor Bolotnikov and me?'

'Alone?'

'Please.'

Yesin excused himself with his colleague and went to the room to which Kaspadov and Professor Bolotnikov had withdrawn.

'Do you need me long?' he asked, after he had closed the door behind himself.

'Please sit down, Mr Yesin,' said Kaspadov. 'We want to show you something. Then you will understand why we have asked you to come.' Kaspadov gave him the same information as he had given to Professor Bolotnikov.

'Mister Yesin, you do remember the robbery of the construction plans of the zwikker. An American spy, Spone, was arrested by you and later released. We have never traced who had had access to our computer, but it is quite certain that two copies have been printed. At present we have the affirmative that a "wild" zwikker is working in Washington and the question arises: was a colleague of Spone involved? I assume that you have had contact with Peasley after the arrest of Spone.'

'Yes, I had. Peasley claimed high and low that Spone had been sent only for testing our security system. But he might have played tricks on us. Probably he has acted on his own and outside the knowledge of President Smith. I can even think of the arguments he has used.'

'That's not necessary,' said Professor Bolotnikov. 'What I would like to know, whether you had additional suspicions and whether you checked those?'

'Only one track, Sir, but that reached a dead-lock. It concerned someone who works at the Western European Union in Strasbourg and who was in Moscow on holiday. Someone of our division, who pursued Spone, recognised this man on the pictures we showed him. We register all tourists who visit our country and he recognised the man twice at the Abbey. Rarely tourists visit the Abbey twice; therefore we have examined this person further on. He is called Jenke Holthus and is the head of an unimportant department in Strasbourg. He stayed in hotel Krymia and remained a week in Moscow. He met several people, but those tracks gave absolutely no result.'

'Would the Western European Union have sent someone to steal the zwikker? But if so why is that thing in Washington?' asked Professor Bolotnikov. 'Do you know something more about that Holthus?'

'We still keep watch him. Someone of our service who works there pays attention to him. Until today nothing particular has been reported. He had only, and which is perhaps peculiar, on 27th October a meeting with Vice-President De Beaufort and was visible angry when he left. He got probably zero on his request for a higher position. Further we only know that he lives very simply, has a family with two children and goes sometimes on holiday to America where he has a get-together with former friends. He was once in an exchange programme of young people. Since his visit to Moscow he has not been there. He is not on the list of moles of the US-security service, and he has conducted no particular telephone conversations.'

'He's probably not the person who entered the zwikker centre, and we have further no other indications. The remaining question is what to do now?'

'May I make proposal, Professor,' said Kaspadov, 'we cannot trace the exact place of that "wild" zwikker with our satellite detection system. It is still too inaccurate. Moreover the satellite passes Washington only once in sixteen days. We should trace the zwikker with a small detection apparatus in Washington itself.'

'Do you have such an apparatus?'

'Yes, we have even several. If we take along two sensors, we can find the zwikker by means of cross detection from several points. You want see one?'

'No, that comes later. I think of another problem. I can hardly inform the Secretary-General of the UN or President Smith, since we made our satellite detection system operational without informing them. We kept silent since our visit in September. They can easily think that we keep other things secret. Is that so, as a matter of fact?'

'In a certain way, yes. We were in September already reasonable far with the anti-zwikker, and haven't reported that to the IZRG.'

'Leave it for the moment that we wanted to trace the "wild" zwikker and destroy it. If indeed the US Security Service is the wrong-doer, then Yesin must wash Peasley his ears and I contact directly President Smith. But first we must have hard proofs.'

'Have you, Nicolai, someone who can be trusted who can go with Kaspadov to Washington? I want Yuri to go in the first place. He has, however, a disadvantage when they notice him. This might get them suspicious, but he is one of the few who knows the problem. They should go as soon as possible to Washington and trace the zwikker. I hope that they take no holidays with Christmas, because then it can last months before we find the zwikker. But, knowing the Americans, they want always everything first thing in the morning, also on Christmas day.'

Yesin deliberated for a moment and said: 'I have no male, but a female who can join Kaspadov. But watch out, she is a very special woman. Are you married?'

Yuri shock its head. He had only once been engaged.

'No? Then all is well. The young lady, whom I propose, is single. For the outside world you can be a couple. She is violist, called Svetlana Shornakova. She practises as much as four hours per day. This you have to endure. If you want, you can call her your sister, but engaged or married is better.'

'What about her manager? Wouldn't that be better? Those other explanations are too transparent. I can then easily quarrel if she doesn't do I say.'

'That's your business, or better, she will help you in that decision.'

'So that's agreed,' said Professor Bolotnikov. 'You go with two zwikker detectors to Washington and report directly to me. I will inform Nicolai. Under no circumstances you are allowed to take action. People should not become suspicious and for any intervention this is a matter of higher authorities. Is that understood?'

Kaspadov nodded. He found it thrilling to play a special agent and hoped only that this Svetlana would be a good colleague. What had troubled him was that violin. As a manager he had to conclude contracts for concerts. And Svetlana should make auditions. Would there be enough time for their detective task? Perhaps he could equip the sensors with an automat, so they could perform 24 hours per day. In this way they could place the equipment quietly in two different hotels. Fortunately the detection apparatus was not large, so that they could take it along in their hand luggage.

The problems with the American customs authorities could be avoided by building the detection in a portable threetel. Threetels were usually transported by almost everyone.

'Are there other points, gentlemen, for which you have lured me to here? I want to see for a short moment the members of the Commission of Supervision, who seem to be today here too. Or is that also a coincidence?'

'Yes, Professor, I had invited them to give your visit an official touch.'

'As from now you can say directly the truth to me, Yuri. But you were right. Concerning the robbery I would have indeed asked more information before coming. Gentlemen, I wish you much success and hope shortly hearing from you.'

The next day Kaspadov got acquainted with Svetlana Shornakova. They had made an appointment for a lunch and he was waiting in a restaurant. Several persons had already come along, but of the female part he did not hope it was Svetlana. When a slim dark person came in, he understood that this had to be Svetlana.

She approached him and asked: 'Are you Yuri Kaspadov?'

Yuri stood up, stuck out his hand and answered: 'Yes, I am. You are Svetlana, I assume.'

'Svetlana Shornakova. Yesin has described you well. He only mentioned a pale face colour.'

Kaspadov blushed even more and he saw Svetlana smiling. 'She's certainly something,' he thought and tried to get himself under control.

'When do we leave, Yuri? I can say Yuri, isn't it? If it is soon, I must arrange a number of things. Tonight I have a concert and I propose you are present. I have understood that you are my international manager. Do not pay attention to my

Russian manager, who will explode if he hears that I will travel with an international manager.'

'Who is he?'

'A stingy. Economically he is a miser. I hope you are a generous manager. I deserve that. Which concerts do you think to arrange in America?'

Yuri was entirely overwhelmed. Would she know why they were going? What did he know about traditional music, to be able to conclude contracts for a concert violist?

It was obviously clear from his face. Svetlana bursted into laughter.

'Now, when do we leave?'

'Eh..., Sunday, here is your ticket. We fly economy class, not to be noticeable. Our supersonic jet leaves Sunday evening 21 December ten hours. We arrive still the same night in Washington. Here you have the reservation for your hotel, which is for reasons, which I will tell you later, at some distance from my hotel.'

'I don't find it at all terrible when I am noticed, hear. In my profession you must be noticeable. Although you might play terrific, one wants something for the eye too. But I'm hungry. Will you order something for me? I'm always hungry before a concert. Tonight I play with the student orchestra of the Moscow Conservatory and on the programme is a violin concert of Anna Tschikofskaya. She's a composer of the previous century. It is a splendid piece of music and it will do you well to come and listen. I know you are a nice scientist, but you lack naturally some culture. And there is nothing above music, isn't it?'

Yuri nodded silly. What has to become of their trip with such a piece of energy? She looked nice, or rather very nice. He called the waiter and ordered everything what Svetlana had chosen. After the first course, the conversation improved and he could at last bring Svetlana to the point to listen on what they had to do in Washington. She asked skilled questions and Yuri understood why Yesin had selected her.

After the dessert Svetlana knew everything about zwikkers and anti-zwikkers. She found she should have one for herself if only to influence cumbersome managers. This said, she looked Yuri deep in his eyes, upon which he blushed again intensely. And not for the last time.

11

Can a real idealist restrain himself?

Strasbourg

After his travel to Moscow Holthus had become a real cumbersome spouse for his wife Astrd. In contrast to his former well-balanced character, he had become peevish and angry to his children at their least being noisy. Astrid suspected in the beginning that it was because his requested promotion had gone wrong, but it was now already a few weeks this way. She loved her blonde spouse and thought what it could be. Was it related to that unexpected official trip to Moscow? Or was another woman involved. But Jenke came always home on time and had no more unexpected official travels. At his office he had an ugly secretary with who he frequently had a dispute. That couldn't be it.

Astrid planned to ask him that evening an explanation. This could not last any more and the children had done nothing to him. On the contrary, they did everything not to make daddy angry.

After dinner, the children had gone up to make their home work and Jenke had withdrawn in his special little library. He was an enthusiast reader of old literature and in former days he could spent complete evenings reading. The last time he simulated reading and that had not escaped to Astrid.

She opened the door of the library, took an old chair and sat down in front of him.

'Jenke you have something. Something very serious. It has nothing to do with your refused promotion. For that are you much too intelligent. It is something else. It is connected with your travel to Moscow and your visit to De Beaufort.'

Jenke gave no sign of approval. His face remained cloudy as it was now already for two months.

'I know you now for eighteen years and I know you better than you know yourself. You are a perspicacious man with ideals, and intelligence and ideals are sometimes difficult to combine. I think you have a conflict for which you don't have a solution. Am I right?'

A vague smile appeared on his face.

'Don't you want to say it to me? Perhaps I can help you? May be just listening would help? You know I can keep secrets even better than you do.'

'Astrid, it is more difficult than everything in the world. I don't know what I must do. It turns around in my head. First it was only anger. Now it's something for which I feel a large responsibility.'

'I'm listening,' she said, afraid that Jenke would not go further.

'It concerns those zwikkers, you know. I have had the construction plans in my hands and I've given them to De Beaufort. I have taken away a copy of someone who had stolen them in Moscow. Someone of our service, who worked for the Americans. And now due to my actions or failures, somewhere in the world zwikkers may be built other than under the frame of the UN. You understand?'

He had a hopeless gaze on his face. 'Astrid, what am I to do?'

Astrid held her breath for a long time. Her Jenke was involved in an international robbery scandal! But why?

'I wanted only that the zwikker stayed in good hands, so that the apparatus would do salutary work. For this reason I haven't made copies of the drawings for myself and sell them with profit. On the contrary I travelled on my own expenses to Moscow to check upon that man who had to steal them for the Americans.'

He told the whole story and how he had got a copy of the plans. When he told about his visit to De Beaufort, he became red of anger.

'You really think that the zwikker can be beneficial for the world?'

'Yes, I still think that. And I had hoped also for our Western European Union. But De Beaufort has put the plans away or has them destroyed.'

'But isn't that fine? Now the developments remain under the management of the UN?'

'I'm not sure. If there is a failure I have to blame myself. I should have persuaded that man in Moscow who had stolen the plans. I should have taken both copies and destroyed them. But he said that the head of the US Security Service, a certain Peasley, had threatened him that his family would not be safe when he didn't deliver the plans.'

'Do you know for certain that he delivered them?'

'Yes, there is no doubt. He has travelled to Washington, I could verify that.'

'And now you blame yourself that the zwikker can be made in the US. But why should the US be less reliable than Russia who has developed that thing?'

'You may be right, but from the fact that there is no news, I suspect that US Security Service is up to something. And about that I can only speculate.'

'Darling, I am glad you have told me this. I think you must do something. I don't mean you must solve it, but simply do something. Don't you have a friend in the US who can check what is going on? I mean one of the friends from the time when you were exchanged with young Americans. From that time you've that American accent in your English.'

Jenke's face cleared up. Just the fact that Astrid supported him and said he must do something, did shake him already up. He had not felt himself this way in months. 'Perhaps you're right. I believe that I know someone. A tree-long and nice fellow. He is called Ben Corward and he seems to work at some development laboratory. I will ring him up tomorrow.'

'No, do not ring. For all money in the world I don't want you to have difficulties. You must simply go to him. When is the following reunion of the exchanged young people? Or the elderly you can better say?'

'Next year at the end of December. But as a matter of fact we can also go privately.'

'Can I go with you?'

'I've no objection, but can you make yourself free?'

'That I can arrange. Mother can come over for one week and Ove and Joanna will find great fun when Granny is there. And certainly after two months to have lived with a muttering father. They asked already whether you were sick. They had really a difficult time, but they were brave.'

'Astrid, I'm ashamed of myself. I have tried to be normal, but I didn't always succeed. I will explain them that there was something very difficult about which I could not speak. Will you inform them that we have to go for a week to the US?'

The children reacted more relieved than concerned when they heard that their parents planned to go travelling. 'And when comes Granny?' With her they had always the greatest fun.'

'Yes, Granny will come, at least I think so. I must still ask her.'

Washington

Thus it happened on Friday morning 19 December, when Ben Corward left his house in Washington to go to his work, that he suddenly was seized by his former friend Jenke Holthus.

'Heaven, Jenke what are you doing here. You scared me. Everything what I expected, was not you here in this street! Do you have a moment to have a drink with me over there?' Ben Corward was really delighted to see his friend. 'They can miss me at my work for an hour or so. Come and tell me why you are here. Is it accidentally that I find you here, or not?'

'Not by chance. I have a lot on my liver and I hope I can participate that with you. I know you work in a development laboratory of the US government, isn't it.'

'And what about that? Has it something to do with my work?'

'Perhaps, but I must speak to someone whom I can trust. I hope it's you.'

'You speak in riddles. Of course you can trust me. Don't we have discussed in former days many an ideal and how we should realize them? We have entrusted each other many a secret.'

'It concerns the zwikker. I know that some construction drawings are in your Security Service. I have seen them. I know who stole them in Moscow and for whom they were intended.'

Ben Corward jumped up with a jolt. 'You mentioned there the zwikker?'

'Sure. You know about it?'

'Yes. To be honest, I work at US Security Service on the zwikker project, but I was never aware that the construction drawings had been stolen by our service.'

'Oof, then it is very good that I speak with you. Your boss Peasley had *a mole* with us in Strasbourg. He works in my department, and was sent to Moscow to steal the construction drawings. But before we go further, I must know whether you still have the same ideals as in former days. Otherwise it has no sense to talk further on.'

'If you aren't different since we met before, you can't expect I've changed, do you? No, I am still the same and that's difficult enough for me. And because of what you told me about Peasley, it becomes still more difficult. Perhaps we must clear up everything, tonight if possible. I must now really go to my work. Where do you stay?'

'In hotel Astoria, not far from here. Can you be there tonight at seven o'clock? Our room number is 555.'

'You say 'us', who is with you?'

'Astrid, and she's informed about everything. She suggested more or less to see you as a friend in whom I could have faith.'

'I think she has done well. I will come, but perhaps I will take along someone. It is my direct chief. He has at present the same moral problems as I've. He is called Ron Sheily. We have no mysteries for each other. Jenke, I'm starting to see the things less black than they were. Until tonight, and count on us.'

He pulled up his long body and hurried away. Jenke remained in the bar for some time, paid for their coffee and walked slowly to his hotel.

He found Astrid still in bed. She had slept late.

'And? Have you seen him?'

'Yes, he will visit us at seven P.M. with somebody else.'

'Oof, who can that be?'

'His direct chief, called Sheily. Ben said that he has problems with the zwikker, too.'

'So, you have discussed already the zwikker?'

'And more, Ben works on it. He is with the US Security Service, but he didn't know that Peasley ordered to steal the plans. He was entirely confused. Tonight we talk further.'

'Have you informed him that I know about all?'

'Yes. Perhaps we can order a diner for four persons on our room. Our conversation might last some hours.'

They descended down and took a late breakfast. They discussed the new information and tried to examine how the different persons on the zwikker check board would react.

Jenke brought forward that one should not underestimate the people in Moscow. They knew perhaps that the plans had been stolen, although they had not reacted officially. That could be explained differently. One possibility was that they shadowed a number of persons who had been in Moscow as a tourist. Jenke had had the last time the feeling that he was followed, but since it was never by the same person, he was not certain of it.

'But since we are speculating Jenke, and if they check on you, they will become very interested when they know that you are here in Washington and will have a meeting with two members of the US Security Service. Let us hope they haven't yet that information.'

'And,' continued Astrid who started to like this kind of speculations, 'what do you think they have done first, after they discovered the robbery of the construction drawings?'

Jenke looked expectantly at his spouse.

'Think my treasure! They are gifted technologists there in Moscow. They will speed up the development of their zwikker detection system. Perhaps they have it ready and using them for finding illegal zwikkers?'

'Astrid, you certainly read the press bulletins well, I must say. You can be right. If that happens, Peasley will be unmasked. Let us hope his zwikkers will be still there.'

'Perhaps your visitors of tonight know more about it.'

'What are we going to do until tonight? What about to go around by car?'

This pleased Astrid and he rented a magmobile. It was a vehicle that could move both on magnetic fields and by itself on wheels. It operated on a compact electrical source. It slid on the motorway, neatly in rank and almost without resistance on magnetic rails. Jenke had programmed a junction which would lead to a demonstration centre on energy, something which he always wanted to visit. The centre was visible from far by some large bulbous antennas. These caught energy that was radiated from space to earth. In space large mirrors of hundred kilometres radius converted sun energy into a newly discovered form of laser radiation which was beamed to the large bulbous antennas. For people in the surroundings there was no danger, at least if one remained at more than five hundred meters distance of the antennas. These form of energy supplied electricity which compensated the decreasing production of fossil fuels.

The number of visitors was not large on this working day, as a result of which they got the occasion to visit everything quietly. Conducted by a kind of walkie-talkie they were informed of the newest progresses and plans for the future. One of the expositions which drew Jenke's attention was which over the new glacial period. It stated that these were farther away than estimated before.

Astrid was especially interested in the models of steel houses. Because of the increased use of plastics and light metals, the iron production had decreased during the last decennia. The metallurgy had searched for other applications and had started to make steel houses in serial. A steel house lasted longer than a stone or wooden house, was transportable, earthquake-resistant and energy poor. Even in the north of Canada they appeared to satisfy. The heat insulation was almost perfect through the use of triple isolated walls.

Jenke and Astrid went next to a historical building, where one could have an old-fashioned lunch. After a delicious long lunch, they took again their magmobile and drove to the nearest highway. There Jenke programmed the junction close the Astoria hotel and quietly lying in their seats they arrived an hour later at the exit. It was five o'clock in the evening and it started already to darken. Snow was predicted and it would become a cold night.

They withdrew in their room, took a warm bath and got dressed to receive their visitors.

'Do you know what I liked with those steel houses?' called Astrid from the bathroom. 'You can take them along if you move. And if you buy one, you can stipulate the model. They do not rust, even not in the neighbourhood of the sea, and they are guaranteed for seventy years. Jenke, do you listen...?'

'Yes, yes, I listen, but I was thinking. If someone taps on the wall this must give a considerable noise. How expensive are these houses?'

'More expensive than stone houses, but that will change if more are sold and the competition increases. You want to live in something like that?'

Jenke who thought of other things, shouted: 'What do you say, where I want to live? Where we live now of course, in Strasbourg.'

The conversation did not go further because the telephone buzzed. Jenke took it and heard someone say on the other side: 'Is this room 555?'

'Yes, that's correct.'

'There are two gentlemen for you in the lobby.'

'Would you ask them to come up? I expect them,' replied Jenke. 'Astrid, our guests are coming up. Are you ready?'

'Almost, only few minutes.'

When. Jenke opened the door he saw his friend, with besides him a man who was two heads shorter. 'Welcome, welcome. You are certainly Mr Sheily?'

'Ron Sheily. Just call me Ron. You are the old friend of Ben, isn't it? He has told me about you.' Astrid, who had left the bathroom, approached them with a: 'Please to meet you gentlemen. Astrid is my name. You are Ben, I presume,' looking high up.

'Not simple to miss me, can I say Astrid?'

'Yes, let us keep it informal. Please take a seat. Later on a diner will be served in our room. I have understood that we have to discuss something serious. Excuse me when I say we, because I'm also involved.'

Ron found her sympathetic and could imagine why Jenke confided in her. He did not do that at home, although it was rather a matter of habit than lack of confidentiality. It was remarkable that he would expose himself tonight to two complete strangers. He pulled nervously with his face and his goatee whipped up and down. Astrid smiled, she was glad that her children were not present. They probably would have maid some remarks. And also of his clothes, which, although clean, seemed to come out straight out of the washing machine.

Jenke asked them what they wanted to drink. 'With or without ice?'

He then told them on the robbery and how he and Astrid had reached the conclusion to consult one of his old friends in Washington. The fact that Ben worked in the zwikker laboratory had been a large surprise.

'But what is your intention with this visit?' asked Ron.

'I wanted to know whether the construction of the zwikker had started and for which aim the zwikkers will be used. Furthermore Astrid suspects that each zwikker which is outside New York or Moscow can be traced. From a press bulletin of a specialist I have learned that the detection of zwikkers is projected by means of satellites. I think also that they know in Moscow about the robbery. The silence about it gives us the presumption that something is up.'

'Will you tell them about our results, or shall I do it?'

Ron hopped with his goatee. 'I will do it. That makes me at least a full accomplice. We have already a number of zwikkers. They have been tested and when the Russians have a sophisticated detection system, it is possible that they have noticed that. I have provisionally informed Peasley of the fact that we have developed some zwikkers. Concerning the detection apparatus and anti-zwikker, which we also have constructed, he isn't yet informed. I have kept this information to myself, since I became suspicious. Peasley has increased my concern at our meeting last week. What I now tell you is a straight violation of my duty to confidentiality and is contrary to the service regulations. After that last meeting with Peasley I have taken the components of one zwikker to my house. I have succeeded to put such an apparatus together. I only lack some parts for the sensor and then it's ready. These I will take along tomorrow.'

The other three looked at him astonished.

'I knew nothing about that,' said Ben, 'that was very clever of you!'

'Clever or not, I've done it. When Peasley wants to use a zwikker for his own aims, there must be at least someone in the world who can stop him. And that can only be done with the same apparatus.'

'Oof, I thought I had problems,' said Jenke, 'but these are of a larger scope.'

'I must confess that I'm just like Ben an idealist and from this point of view I've committed this house robbery. Perhaps it is in the future necessary that somewhere in the world we have to intervene with a zwikker when the UN for example cannot act fast enough, I.......'

He hesitated as if he realised only now what he had said. What could go wrong by his fault...., he couldn't overlook everything as a technician...?

'Have you clear plans with that thing?' asked Astrid.

'No, as I said only to prevent a catastrophe when Peasley will use it for his own aim. And..., one never knows...'

'We must arrange all things as judicious people,' said Ben. 'The house robbery should definitely not become around. Ron, you have to put this zwikker well away and certainly not let it working. Otherwise the US Security Service, or rather we ourselves will trace him. Or, the Russians will do that for us. It would give an international scandal which goes far above our objectives. Your zwikker can only be used when we three are entirely convinced. I count Jenke and Astrid as one person. And if so, we can use the zwikker only for some minutes. Furthermore, if it is possible, never at the same place.

'And then still something. I do not think that one of our three wants to jeopardize his job. We must also live and the one, who is the cause of our behaviour, is Peasley. From the tales of Jenke I have understood that we cannot count on the Western European Union. We will be on ourselves what concerns our zwikker.'

Everyone was silent, only Ron's goatee hopped up and down.

During this silence, the buzzer of the telephone went and everyone jumped up. Their consciousness played already tricks on them.

It was someone from the restaurant: 'Can we serve your diner?' Astrid answered affirmative and somewhat later they sat celebrating their meeting with a complete diner and a gentle wine. This "celebration" alternated from their ideals to worrisome shivers concerning the consequences for their families. Peasley and De Beaufort would not hesitate to fire them straight away and this would be the mildest what they could expect. Arrest for high treason would be more obvious.

They separated by midnight. Ben went together with Ron in the same direction. They lived close to each other. Both spouses were waiting for them, as they had done frequently the last months. Why their men did so much overtime they didn't know exactly. It had to do with some secret project.

Strange enough Kaspadov stayed in the same hotel Astoria. If he had had a zwikker for himself, he could have measured a particular consciousness field in room 555. He had brought Svetlana first to another hotel and had promised her to install the next morning the detection equipment. He had not dined with her; he could do that later frequently enough.

He had already installed his own detection equipment in his room 455, but so far no zwikker was operating in Washington. They must have patience and Kaspadov was not unhappy with some relief. He had worked very hard the last months and he liked the prospect to be with Svetlana very much. She was clever and passionately. He had to go to some agencies tomorrow to conclude a contract for concerts. He had become deeply impressed by her last concert. Her violin had sung and complained and although he did not admire modern music he had been moved by the music of the composer Tschikofskaya. Or had he paid too much attention to Svetlana?

He looked at the last news on the American threetel, but except some ordinary assassinations, and international problems there was nothing that interested him. Would Svetlana fee something for him? He should have liked to measure it. Why not making a personal zwikker? He forgot that human speech existed and that he simply could ask her. But would she give an answer? It was sometimes difficult to understand girls for a man.

His telephone ringed and got Svetlana on the small screen. 'You are not yet asleep, my hero. Have you already written?' They had agreed to call the "detection apparatus" their "writer" and "measuring" as "letter". Thus they ran less dangers if they spoke in public or over the telephone.

'Yes, but I don't know what to write.'

'Then think what you will say tomorrow to the theatre director. I have practised tonight for an hour to train my fingers. Shall I play something?'

'No, no, do you know what time it is? Tomorrow at nine o'clock I've to speak with you about those concerts. I'll come in the breakfast room of your hotel, because as I already said, I will be busy. I was just thinking of you.'

'Don't do that Yuri, don't! You can better count music agencies, and then you will fall asleep rapidly. If I fill up your mind you will get only nightmares, because I keep talking and play the violin.' On the screen she smiled to him, waved with her hand and broke the connection.

Yuri sighed: 'What a girl. I don't think I have the courage to say to her that I like her. He shook his head as if he wanted to relief himself from all thoughts, took a shave and shower and crawled under the blankets. In his dreams talking zwikkers and singing violins alternated with each other. But since he had followed in former days a dream course, he managed to calm down the zwikkers and violins and sunk in a deep sleep.

12

Where a man stumbles, a woman can find a solution.

Washington

Peasley entered as usually at half past nine his office. He never came earlier. He preferred working late at night and was fond on a quiet morning. Seeing again his office and the large leather chair did him good. He felt himself better and put his feet on the desk. He was not impressed by the fact that psychologists qualified "the feet-on-desk-syndrome" as an inferiority complex. As if one was looking for protection by the desk and make a larger impression on visitors. That last was, however, exactly his intention.

He caught the bulletins which his secretary had put on the desk and started his computer. He had programmed his computer so that in the morning automatically news on persons who interested him was given. In the first place those persons, which had to do with the zwikker.

Peasley examined the list which was projected on the screen. In the first category there were not many travellers and he already wanted to switch over to another category when he saw the name Kaspadov. Entering the US on 21 December, staying in hotel Astoria, profession music manager. Peasley shot up. Yuri Kaspadov from Moscow, clearly as music manager.

There could be of course several Kaspadov's, but not several Yuri Kaspadov's. And what was he doing in Washington? Yuri Kaspadov of the Moscow zwikker centre had been repeatedly in the New York for discussions with the group in the Central Hospital. But now he had come as a music manager to Washington! Peasley's neck hairs stood right up and he resembled even more to a pig then usual. A very suspicious pig.

He pressed the intercom and called: 'Harry, call immediately Spone.'

'O.K. Sir,' answered his secretary, 'I'll see if he is in the house.'

'Not seeing, track him down and let him come here immediately.' Peasley interrupted the intercom and muttered: 'That rot always "I will look if he is in house"...'

He checked the list on the screen once more, but no one else had come in the US. What had Kaspadov to do in Washington under the pretext of being a music manager? What does he have to manage? Musicians or music books? He checked the list again, but unfortunately the computer was not programmed on musicians.

The intercom hummed. 'Sir, I have traced Spone, he can be with you in five minutes. You need his file?'

'No, I can do that myself, just let him in strait away.'

Harry Snowden sighed. 'The big boss has again one of his tempers. When he came in he was radiating, even muttered good morning. Would there be something special between the bulletins this morning?'

Spone came running in after exactly four minutes. He was nervous after his last adventure in Moscow and Uzbekistan.

'Go in quickly, Mike,' said the secretary, 'the boss has already warmed up.'

Spone knocked and blew in after a loud 'Come in!' of Peasley.

'Spone, you can possibly repair something of your disaster in Moscow and Uzbekistan. You know Kaspadov by face and I want to know whether the Kaspadov in hotel Astoria is our zwikker man. He has arrived yesterday and has given as profession "music manager". Go straight away to hotel Astoria and check if he is our Kaspadov. Be careful, he definitely should not perceive that we are after him.'

'Is this all?'

'Are you deaf? I don't care about all. Immediately, I say! I expect your report within an hour. Scram it!'

Spone stammered 'O.K. Sir' and 'Immediately Sir' and disappeared as quickly as possible. At the secretariat he asked if they knew more about it, but Harry Snowden, being wise for a long time, ignored him.

Spone ran outside. He slipped almost on the street because there was still some snow. He drew his coat up to his ears and took the subway. With this snow it was faster than by taxi. Even the most ultramodern magmobile remained blocked if there lay a package of snow. It was against ten fifteen when he arrived in hotel Astoria. Mike shook the snow from his coat and went to the counter.

'I have an appointment with Mr Kaspadov. You can put me through?'

The man behind the counter pushed the telephone to him and said: 'Number 455, Sir.'

Spone disconnected his own telephone camera so that he would only get the picture of Kaspadov without being seen. The telephone sounded, but nobody responded.'

'Is Mr Kaspadov in?'

'Can't say Sir. There are three hundred guests and everyone keeps his code card for opening his door. Must I hand over a message?'

'No, that's not necessary.' He stood in conflict what to do now. Phone Peasley that Kaspadov was out? He would explode. Spone decided to act on his own initiative. He went to the lift to the fourth floor. Perhaps Kaspadov's room was cleaned at present.

He had if necessary a "decoder for code cards", but that he could use only in extreme cases. In fact one was indeed busy with the rooms. He entered one where he heard sounds.

'Help, you scared me off,' exclaimed a young lady. 'Mister, what are you doing here? This is not your room, because it was emptied this morning.'

'Sorry Miss, I only was looking for you because I have left my code card in my room. Now I can go to the counter downstairs, but you can help me more rapidly. I'm in a hurry because someone is waiting for me in the lobby. Here are twenty dollars for you.'

The room maid hesitated, but would not let run a tip as large as that.

'Well, but I have to go with you inside. That's regulation.'

She went with Spone to room 455 and let him in. Spone ran assured to the cupboard and did as if he taken the card from a coat. He put this so-called card in his own pocket.'

'You found him?'

'Yes, thanks. You see how it snows outside, don't you find too? It gave him just the occasion to look around. There was apparently nothing particular in the room, only two threetels instead of one. A small one and a large one, probably from the hotel. He talked still about the snow and ran to the desk to look at the screen of the small threetel What he saw was a flash-light which was accompanied by two data on the screen, one in kilometres and one in degrees.

'Are you ready Mister, I must go to my work. Would you please come?'

Spone had seen enough and went outside. 'Thank you very much Miss, now I come at least on time for my appointment. He walked to the hall and called Peasley's number on his pulse telephone.

Spone got him directly on the line as if he was waiting for him. 'And Spone, was it Kaspadov?'

'I don't know yet Sir.'

'What, you don't know yet and you dare to ring me. Duffer, I should never have hired you.'

Spone's face became red. He heard and saw on the small screen Peasley razing, but what he said went beyond him. Spone mastered himself and when Peasley had to breathe for a moment, he said: Sir, there is an apparatus in Kaspadov's room that according to me has nothing to do with music. It measures kilometres and degrees.'

'What do you say there? Couldn't you have said that before? Come immediately here and explain it to me.' Peasley had cut the connection.

Spone stuck out his tongue against the empty screen and left the hotel.

It took him more time than the outward journey, because snowing had persevered. The big boss waited for him with someone who wore an untidy costume and a goatee.

'Tell and be to the point. This is Ron Sheily, head of our special laboratory. What did you see in the room of Kaspadov?'

Since Peasley hadn't informed Sheily, Sheily reacted as a frightened grasshopper. His nose resembled more than usually a hook. 'Kaspadov?

'Sheily, don't interrupt. Listen only. Now Spone, come up with your story.'

Spone found that not Peasley but he himself had been interrupted. 'I've looked into the room of Kaspadov. There were two threetels, a small one and large one. On the screen of small there appeared always two data and a small flashlight. One gave kilometres and the other degrees. The numbers were always the same.'

'Have you noted these numbers?'

'Yes, one was two and a half kilometre and the number in degrees was twenty five.'

'Do you know what that means?' asked Peasley to Sheily.

'If you think what think, we must put the distance on a map. You have one at hand?'

Peasley pressed the intercom and called: 'Bring immediately the map of Washington.'

Snowden got so many commands per day that he became no longer cold or warm. He took at his ease the map of Washington from the cupboard and brought it to his boss.

'Here you have a ruler and look what is so interesting at two and a half kilometres from hotel Astoria.

'That are we.'

'Why we, what do you mean with we?'

'Look for yourself, hotel Astoria is at two and a half kilometre from us and we are with respect to the north of this hotel at twenty five degrees east. What is on room 455 is a zwikker detection apparatus and it is targeted on us. We are at this moment working with a zwikker and the radiation which it transmits is picked up by that apparatus.'

Peasley became pale. Truth dawned on him. The Russians had tracked him down and it would last not long before the complete world would know it. But his hesitation did not last long. It was known that he could react rapidly.

'Stop immediately your zwikker and don't start it again, under absolutely no condition, you hear!'

'And Spone, you return as fast as possible to hotel Astoria and take three fellows along with you. Here you have a warrant and you return with that detection apparatus. Furthermore you leave behind someone who follows Kaspadov if he returns. And do it rapidly, this is a command! He took from his desk a paper, signed it and gave it to Spone.'

Spone left the office, called for three US-security agents and left with a service mobil.

At the hotel he showed the warrant at the counter and asked for a cursor code card. With this card they rushed up. On the fourth floor the room maid was still busy. Spone pressed the code card in the split and the door of 455 opened.

Yuri Kaspadov, who had returned two minutes ago stood in front of his small threetel. The screen gave absolutely no information and he was just busy to examine if the apparatus had registered something, when Spone and the three agents stormed in.

'What.., what is... this,' stammered Kaspadov. 'How dare you invade my room?'

Spone who recognised Kaspadov, said: 'US Security Service, please follow us to our head office. Orders of the big boss. Here is the written order. You can be silent but you should not oppose us. You can catch your coat and follow these agents?'

Yuri was taken by his arms and carried away from the room. He was too staggered to protest. He recovered only somewhat later when he was put in the service mobil. Only then he realised that the man who had addressed him was not present.

Spone had stayed behind and placed the small threetel in the suitcase which he had found in the room. With a taxi he returned to the head office where he presented himself again to Peasley.

In the office Spone unpacked proudly the suitcase and placed the small threetel on Peasley's desk. 'This is the small apparatus about which I told you. It still worked, but it gave no more numbers or the screen. At the same time we have arrested Mr Kaspadov. He had returned and was busy with this apparatus.'

Peasley looked back at the apparatus and at Spone. That the apparatus was confiscated was lucky, but that Kaspadov had been arrested, was a setback. He had hoped that Kaspadov hadn't yet returned. This was bad luck. It would have been easier when Kaspadov would have thought that his apparatus had been stolen.

Peasley muttered something of a compliment and said: 'Spone, so far O.K. Put Kaspadov for the moment in a guest room. He's not allowed to speak with anybody or by telephone, but he can have furthermore everything what he wants. I don't need you for the moment.'

When Spone had left he rang Sheily.

'Is this a detector?' he asked.

'We must dismantle it carefully. It resembles a threetel, but that can be camouflage. If you approve I will take it along for further investigation.'

'Do that and give me soon your results. And test it first if it measures something when you start a zwikker. But very shortly, because there might be more of these things in Washington.'

Sheily took along the equipment under his arm. Arrived in his laboratory he called Ben Corward to join him.

'Peasley has got a zwikker detection apparatus and you never guess where.'

'A detector, today?'

'Yes, this morning and in room 455 of hotel Astoria. Kaspadov was there with this equipment and Peasley has confiscated it and arrested him.'

'Oh dear..., is that real? Told Peasley you that?'

'No, I ran into Spone.'

'And this Kaspadov was only one floor lower than the room where we discussed with Holthus and his wife?'

'This is probably a detector. Disguised as a threetel. Would you please start that zwikker there when I start this equipment? Thus we can measure whether it really works.'

Sheily connected the sensor and the screen lighted up. When the zwikker started with the usual sound proceeding to supersonic, he saw on the threetel screen a flashing light and two numbers. The upper number indicated thirty degrees, the other gave two meters.

'It's indeed a zwikker detector, no doubt possible. And Kaspadov has traced us this morning. We have caught the detector just in time. Thus, the Russians know that there is a zwikker in Washington. I suppose they came here to find it. I think we must discuss this directly with Holthus before he flies back to Europe. This afternoon. Can you give him a ring? Or better, pass by there. I will dismantle this apparatus in the laboratory so that Peasley must wait for an answer. I will inform him in an hour time.'

When after an hour Sheily informed Peasley that the apparatus was indeed detector, and wanted leave his office, Peasley beckoned him to stay. On his thick face he showed changing expressions. He was probably struggling with a personal conflict.

'How many zwikkers have you now?'

'Three working and two in production.'

'Three, or rather five,' Peasley looked at the sealing. 'In order to prevent that we would soon be without, you better bring one to me. I will store it personally. The other four we can hand in if we are asked for. Although there are still no rules, at least I have had still no instructions, I can imagine that new developed and traced zwikkers must be made available to the UN. And obviously Kaspadov has tracked us down.'

Peasley hesitated. 'Perhaps we were just in time and the bosses of Kaspadov aren't informed yet that he had detected us. Is it possible to trace zwikkers from space?'

'I don't know for certain, but with hundred times more sensitive detector than ours..., I think yes.' Sheily remained on the careful side. He was, however, convinced that it was possible.

As if Peasley had read it on his face he spoke: 'So it is and perhaps therefore they came to Washington. Those sneakers have kept back information for the UN and are busy on their own. Sheily bring me one working zwikker, including the instruction notebook and the construction plans. Store the rest well. We will provisionally stop working with the zwikker, only on detection equipment and anti-zwikkers. How far are you with the last?'

'As I already reported to you, we have progressed, but it is not yet finalised.'

'Make it an emergency. It is important that we have them rapidly. The Russians seem to progress very fast.'

Peasley beckoned Sheily to leave, but called him after: 'Bring straight away that zwikker!'

Yuri looked taken aback around. Everything had gone so rapidly. He had hardly realised he had been caught. The room was luxurious, but the heavy doors were locked. As a matter of fact, the windows were not real. They were bogus windows from which diffuse light entered the room.

He thought of Svetlana who he had visited that morning in here hotel. They had had together a cosy breakfast and afterwards he had installed her detector. To their joy the thing gave a result with a distance in kilometres and an angle in degrees. On a map, given by the hotel, they had designed a circle with the hotel as centre. Since the angle on the screen vibrated, they could determine only a segment towards the centre of Washington.

Nicolai Yesin had told them to pay particular attention to the US Security building in the centre. Svetlana applauded when Yuri pointed to the place where the building was located. It was indeed in the sector.

'They use a zwikker, Yuri, and we have found it. What will do you now?'

'I will warn Professor Bolotnikov, but before I ring him, I will return to my hotel and compare my results with your's. Then we are entirely certain and don't have to come closer. Although...., perhaps..., you never know? Perhaps they have more than one zwikker operating. I will hurry and check what my sensor has registered this morning.'

'Yes, do that Yuri, do you want me to come along?'

'No, you better stay here. I'm back in an hour. Afterwards we have enough time for managing your music. I have collected a number of addresses which we can contact this afternoon.'

Yuri looked Svetlana in her eyes, blushed slightly and disappeared.

Svetlana made a large pirouette. She felt herself floating. She found Yuri a nice fellow, nicer than anyone she had met in years. It would be marvellous to stay here for a couple weeks. And give a concert, although the preparation was very short. But you never know. She caught her violin and played the adagio from a violin concert of Mozart to which she was very much attached.

She was so much taken by her music that she forgot the time. After Mozart followed Bach and Tsjaikowski. She felt herself happy. When there was a knock on the door, she realised that Yuri should have returned for a long time. She opened the door to welcome him, but it was the room maid who asked if she could do her room. Slightly disappointed she said O.K. She decided to go to the bar.

When she returned after half an hour, Yuri had still not appeared. 'Something must have happened to him. She caught the telephone and called the number of the Astoria hotel. She was put through, but nobody replied. Yuri was certainly underway. Svetlana waited another hour and rang again Yuri's hotel, but without result. She wondered what to do. Ring Professor Bolotnikov? Only because Yuri was late? She looked at on her watch. Yuri was already three hours over time. Something must have happened.

She took a decision. Waiting longer had no sense. She caught the telephone, tapped the number which Professor Bolotnikov had given her together with his code. The line remained quiet for a minute when she saw the face of Professor Bolotnikov appearing.

'Is something the matter, Svetlana?'

'Yuri has disappeared, Professor. He should have returned for three hours and he is still not in. What must I do?'

'Tell me first what has happened.' His voice sounded quiet.

Svetlana told everything as from the moment they in the landed US up to the moment that Yuri had gone with the promise to return in an hour. 'He wanted to compare his detector with mine. His detector was on automatic.'

'And what did it measure?'

An operative zwikker in the immediacy of the US Security building. With Yuri's we would have been able measure precisely whether it was this building. But he has still not returned! What must I do?'

'Svetlana, remain quiet. Go under no circumstances to the Astoria hotel of Yuri. One disappeared person is enough. I will send a confidence man to the Astoria hotel and I will ring you tonight what he found out. Keep quiet and take care that you stay in your room. You understand?'

Svetlana nodded, upon which Professor Bolotnikov broke the connection. Not long afterwards he caught Nicolai Yesin. 'I suspect that someone has captured our Kaspadov. He has disappeared, at least he has not returned at Svetlana. He is already three hours over time. Have you a confidence man in Washington who can check at the Astoria hotel? And let him check whether Kaspadov's detection apparatus is still in his room. Svetlana told me it is room 455.'

'I know someone who can do this job.' Yesin needed only three words to act. Within ten minutes he had his fellow in Washington on the line. It was a minor actor who lived from small roles in plays and musicals. The supplementary earnings from the Russian Security Service were very welcome. Mario Tkachov was his name.

'Are you there Mario? I need an information. You must penetrate unnoticed the hotel Astoria and ask what has happened with the guest Kaspadov of room 455. Furthermore you must examine this room, whether there is a small threetel. There will be the common large one, just like in all other rooms. I must know whether that small one is still there. If so you take it along with you. Find this out and call me directly after your visit.'

'O.K. boss, still more of your service?'

'No, this is everything and also enough for you. Don't get caught! It is of the largest importance to us. And work fast.' Yesin broke off.

'A job of nothing,' said Mario to himself. 'I will show what kind of a professional I am. A small threetel. These have to be repaired permanently. And I am expert in repairing threetels.'

He went to his workshop where he had stored his disguises. He took a blue overall showing on its back the initials IHES, meaning International Help Expert Services. He was proud of this company. With his old magmobile and a tool kit he sped to the Astoria hotel There he went through the service entrance and moved straight to the fourth floor. Each floor of the Astoria hotel had its own room maids and one should at this time of the afternoon be in the linen room.

He opened the door and said: 'Beauty, I have been rung up for an emergency in room 455. The guest wants a small threetel to be repaired.'

'Jonny, do you have an authorisation from the counter?'

'Of course dear, what do you think? They called me up here, and I have still more to do today. Can you let me in. And stay with me so you can see how an expert works. Look here IHES, international help expert services. If you have troubles at home..., call me and everything will be repaired.'

'Say, you feel yourself quite a fellow. And because of that I must open again the door of room 455. First the guest this morning who claimed that he had forgotten his code card in his room. Afterwards four big fellows with a cursor card. They entered and five man left. All in a hurry. And now you. Well, come on, I'll help you. Perhaps are you nicer.'

Mario understood the hint and pressed ten dollars in her hand. She took him to room 455. Entering he saw no small threetel. 'Where's that thing which I must repair?'

'Don't ask me, man. If I have to reply to all questions people ask me, I will never finish my work. And the businesses of the guests don't concern me. I can tell you stories of which you would shiver.'

'Do you know perhaps of one of the four man has taken something with him? A suitcase or some box?'

'Let me reflect...., thank you,' she said, after receiving a second tennar. 'Let me think, first four man left, later the fifth came with a small suitcase. Certainly also international repairers, just like you. You have to find out for yourself. If that thing is no longer her, you have nothing to do here anymore. And take along your tool kit.'

Mario had heard and seen enough, he gave the room maid a chuck under her chin and avoided skilfully the slap which she had intended for him. 'Deary, a lot of thanks. If you have something to repair, call me then. Here is my business card.'

'Keep your card, I know what your kind of fellows want to repair, not me! There is the elevator and please stumble over the threshold.'

Mario bumped almost against a long man who came bended around the corner. 'Sorry, Sir. I didn't see you arriving. That room maid there, doesn't want to go out with me tonight. She's very angry this moment.'

Mario disappeared and Corward, because it was him, looked at him astonished. After the conversation with Ron Sheily, they had visited immediately the Holthusen and had reached the conclusion that he had to examine room 455. There was always a chance that the agents of the US Security Service had not examined the room thoroughly and that he could find an indication of interest.

He walked up to the room maid and showed his US Security card. 'Can you let me in room 455, Miss?'

'Good lord, you too? The complete world wants to be in room 455. No, Mister I've enough of it.'

Ben went in front of her, so she could not move anymore. 'Miss, I have nothing to do with the complete world. I can let you arrest if you don't collaborate with an agent of the US Security Service in function. Would you please open now room 455?'

'Yes, yes..., you don't have to threaten me. That repairer, in which you bumped, gave me at least ten dollars.'

'You mean your lover?'

'For death not my lover, that Jonny. A small threetel repairer he called himself, but it was no longer there.'

Ben pushed twenty dollars in her hand and the face of room maid brightened up. 'That's better than threatening, Mister. Why everyone is interested in room 455, beats me. You can go in, but you can't take anything along, security service or not. Otherwise I call the internal service of the hotel.'

'I won't take anything, believe me!' Ben entered quietly the room. There was indeed no small threetel, but the luggage of Kaspadov was still there. Beside a snug corner stood a large and a small trunk. Some costumes hung in the cupboard and the shirt had been arranged neatly. Kaspadov had therefore counted on a long stay. At the telephone he saw a note-book with one described sheet. There was one number on it. He tore the sheet and put it in his pocket. In the trunks he found nothing particular. On the table were some folders of some concert theatres. He took these along too, searched under the bed, but found nothing furthermore. Predecessors had taken along probably all personal papers of Kaspadov.

'Miss, I am ready and I thank you warmly in name of the security service for your eagerness.' Ben nodded slightly to the room maid who had watched all his actions. 'Here...., this hundred dollar bill is for you, provided you won't tell anybody about my visit. Nobody, also not other people of US-Security Service. You understand?' He gave her a wink.

The room maid winked back and seized rapidly the hundred dollar bill. More like that were welcome. She could make an exhibition of room 455. 'You're of course not of the US Security Service.'

'People should not underestimate you, Miss. And all the more reason for you to tell nobody about my visit,' he added grimly. 'Otherwise we know to find you, count on that.'

'I've already forgotten you, Mister. I've never seen you and you've never been here.' She looked at him worried, but recovered quickly. 'And scram it that's better for my memory.'

Ben left rapidly, went to the elevator and pressed on number five. Arrived on the fifth floor he ran to room 555 and knocked. Holthus who had waited for him, opened the door. 'You've found something?'

'Not much, but perhaps enough. A number of folders and a telephone number. What we will do, ring this number?'

'Not so fast, let us discuss it first. You can trace that phone number?'

'If it's not a secret number, sure. On the telephone apparatus is a computer system.' He tapped some numbers a found the subscriber of the number. It was hotel President.

'So Kaspadov has rung someone in hotel President or intended to do so.'

Astrid, who had listened, said: 'What kind of folders did you take? Can you show me?' After regarding them she said: 'These folders were not in our room layers, thus they must be of Kaspadov. Perhaps he wanted to call someone in that hotel about music. If you ask me we must look for a lady of approximately his own age, which is musical and stays in hotel President. This is a beautiful job for me, boys. You stay here, I mean Jenke and I go to hotel President. I will find out. If she is a Russian too, then I have her in a short time. And you Ben, better disappear, because that room maid must no longer see you. She will of course betray you, something like that is too interesting for her.'

Ben and Astrid left together the room and separately hotel. Astrid took the underground to hotel President.

Arrived there she ran straight to the counter. 'Sir, I am of a Danish theatre group who expect a number of Russian colleagues. They are musician. Have some of them already arrived?'

'What's the name of that group?'

'Tsjaikowski, the concert group Tsjaikowski. But I'm not certain whether all of them will stay with you. It's possible your hotel is too expensive and that only some soloists are in your hotel.'

'That will be difficult to check, do you the names of those soloists? On the name Tsjaikowski no reservations have been made?'

'But the phone number on this paper is nevertheless yours?' asked Astrid, which saw herself placed for a wall of hotel clerks.

A colleague of the man at the counter saw how worried Astrid was, gave her a wink and said. 'George, this morning there was a complaint about that lady Svetlana Shornakova on the third floor. She played violin yesterday evening and this morning so that some guests next to her became enraged. Did you check with her?'

'Yes, yes..., Jack, I will come to you in a moment. I have first to help this lady, or rather not help. I'm really sorry Miss, we can't pass on names of guests. I'm terribly sorry.'

Astrid, delighted by the intervention of colleague Jack asked: 'Can I ring somewhere? My pulse telephone is out of order.'

'There in the corner, Miss. For external calls you must tap first the zero.'

Astrid thanked them and went to the telephone cabin. She tapped the hotel central and asked: 'Can you put me through to Miss Shornakova?'

'A moment please,' she heard and a click. Nobody replied but on the screen appeared a young woman.

'Bingo,' thought Astrid, 'with such a woman I can imagine Kaspadov would like to go to concerts.'

Aloud she asked: 'Are you Svetlana Shornakova? I would appreciate to speak with you, is that possible?'

'About what, I don't know you.'

'Indeed, but it concerns Yuri Kaspadov.'

Svetlana remembered that Professor Bolotnikov would react rapidly, that perhaps this was the person he had contacted. But hadn't he said he would ring her first? Or had something happened which had modified his plans. She hesitated. The woman looked reliably and she had spent the last hours in large tension. What to do? Why not see her. Nobody knew Yuri in Washington?

'Please come to my room, I'm in room 326.'

Astrid broke the connection and went to the elevator. The lift boy brought her to the third floor and on the left she found number 326. She knocked and was immediately let in. Svetlana was smaller than she had suspected from the screen. She was particularly nervous.

'Astrid Holthus,' she presented herself, 'and you are Svetlana Shornakova.'

'What do you know of Yuri.... you mentioned his name? What has happened?'

'She's in love with Yuri,' thought Astrid and continued aloud: 'Yuri has been taken this morning into custody by US Security Service along with his zwikker detector, exactly such equipment as you have there.' Astrid had gambled on good luck and from the response of Svetlana it became clear that she had made shot in the rose.

'Svetlana, if I may call you. I think that I have to make a declaration. In Washington nobody knows you are here, at least not as zwikker detector expert. It is purely chance that we know you are here too. I myself my husband, Jenke Holthus are not from the US Security Service. But we know they had developed zwikker.'

'How do you know this all..., how's that possible? There seems to be more going on than I've been told.'

'We, my husband and I know that you are here for a good aim. Tracing a so-called wild zwikker. By chance Yuri Kaspadov was in our hotel and people of US Security Service, dear friends of us, have informed us. One of those friends has searched Yuri's room and he found the number of your hotel and some folders concerning concerts, whereupon I've found you. To which concerts you wanted to go?'

'I'm myself violinist and Yuri is my manager. He wanted really to act as a manager and I've practised my repertoire yesterday evening and this morning.'

'Fortunately you did that, because there was a complaint at the counter which I heard accidentally so that I gambled that you were a friend of Yuri.' She saw Svetlana wrestling with her feelings and with her consciousness. Could she trust this visitor? She was certainly not allowed to speak about her mandate.

'I will be entirely open to you,' said Astrid. 'It started this year in September....' she told her everything what she knew about the visit of her husband to Moscow, the vexation with the President of the Western European Union and the meetings which they had with and Ron Sheily and Ben Corward. At least, almost everything. She did not mention that Ron at home had put together a zwikker.

'So you see, we are in a plot. There is something dangerous going on at the US Security Service and we wanted to do something about it, not knowing you were also on the track. How did you know that there were zwikkers in Washington?'

'Satellite detecting. Russia has placed a detector in a satellite and this one located an operating zwikker somewhere near Washington. Yuri and I have been sent out to detect where that was. And now after one day Yuri has been caught. What will they do with him?'

'I can find that out for you because our two friends work at US Security Service. One is even the head of the department zwikker research. I will report to you shortly what will happen with Yuri. At least if my friends can find out. You have already informed Moscow?'

'Yes, I have. I expect tonight a reply from them. You don't betray me, would you? I wouldn't know what to do. Do I have to dispose of that detector? What do you think?'

'I suppose Moscow will tell you, and don't worry, I will not betray you, certainly not, because we would then betray ourselves too. It's all very exiting, isn't it, I like it.' And after a moment of silence she said when she saw the perplexed face of Svetlana: 'They won't harm Yuri, sure. He will return and then you can fold him in your arms.'

Svetlana blushed slightly. 'But we are only temporary colleagues for this task, there is nothing more.'

'That will change, mind my words. I've seen a lot of young girls who were in love and you are one of them. Or doesn't Yuri care for women?'

'No..., no, or yes.., yes. He seems to like me.'

'Now then, there is nothing wrong, everything will become all right and perhaps sooner than you think.' The charisma of Astrid had made impression on Svetlana and she clearly calmed down.

'Thank you, you've comforted me quite a bit. It's easier to play music than to act as a spy. I won't do it again, spying at least.'

Astrid took farewell, embraced Svetlana and left the hotel. She hurried to her husband to tell him everything.

In the meantime Mario had telephoned with Yesin and informed him that both Yuri and the zwikker had been taken along by four men who were obviously of some police force.

Nicolai Yesin knew enough and rang Professor Bolotnikov to notify him that most likely Peasley had confiscated the detector and arrested Kaspadov. He asked instructions what to do, but Professor Bolotnikov wanted first to hold some consultations. In particular Peasley should not suspect anything. Professor Bolotnikov rang Svetlana and notified her that a certain Mario would come to pick up the detector and that she would be free, as long as she did not search for Kaspadov. That was a matter of higher level. He recommended her to give a concert and return afterwards. She could continue to celebrate Christmas and New Year at his cost.

Svetlana wanted to tell him of the visit of Astrid, but Bolotnikov had already broken the connection.

13

The cause of a world crisis can be small. Also what to do about it.

Washington

President Smith, a real beefy American, former football player, lay relaxed in his chair. He looked outside to the snow. Joan knitted a sock. She liked that. 'Even a President shouldn't have cold feet,' was her motto. On the second day of Christmas it was quiet. John had told his secretary that for nothing and for nobody he was available. They wanted to spend the second Christmas day together. For the evening they had asked the Secretary-General of the UN and her spouse to dine. John did not really like Dolores, but he knew he should count with her. The diner gave the occasion to discuss a number of items.

John knew that a single speech and simultaneous indoctrination with a zwikker would have a tremendous success, but could you do that? John had followed the developments around the zwikker with interest. The medical experiments had started at last to give some fruitful results, in particular curing dangerous extremists.

The tests by means of threetel retransmissions had confirmed Raskadov's action in Uzbekistan.

'This must be examined furthermore,' thought John, 'because if there is something what politicians want, is a machine to reinforce their personal charisma.'

'John, what you are dreaming? Where are you with your mind? I hear you sigh now and then. What's wrong?'

'Nothing really, darling, or better said, yes darling. I was thinking on the possible applications of the zwikker, in particular against extremists. Hell, and you're just knitting a sock!!'

'Don't bark on me, John. You've made me drop a stitch. I need every ounce of concentration for the turning of the heel.'

'For what poor fellow is that sock this time?'

'For you, dear. But how far are they with that zwikker? I haven't heard you speaking about it the last time. They are nevertheless busy in New York and Moscow, isn't it?'

'Yes that's correct and it seems to progress. That equipment can be used usefully, in spite of the fact that it's immensely dangerous. For this reason I'm glad they have succeeded in Moscow to build zwikker detectors. The anti-zwikkers are still in a test stage.'

'What do you mean with anti-zwikkers?'

'A thing that can destroy a zwikker at distance.'

'My god, this might give war if you don't watch out. You told me that the current science is able to make zwikkers anytime? The complete world might go zwikkering and anti-zwikkering. And the question is who's the fastest. Or am I exaggerating?'

'I don't know for sure. But the silence around it is distressing. I can hardly imagine that small and large countries are waiting quietly what the results will be of the UN research in Moscow and New York. Everyone is kept informed, but a State has also the duty to protect itself against undesirable zwikkers.'

'You do that too? Developing zwikkers, zwikker detectors and anti-zwikkers?'

'Yes that has been permitted. At least if we do it openly. The question is whether other countries do that too. The seduction is great to have the zwikker and anti-zwikker for yourself.'

'Now I understand why you have invited Dolores and Jorge for the diner. You want to speak with her about this, isn't it?'

'You're right. Each conversation on the usefulness of the zwikker leads also to the dangerous sides of that thing. I would have welcomed that the thing did not exist.'

The butler came in and said: 'Mister President, at the gate stands Mister Peasley, the head of your security service and he wants to speak to you urgently.'

'And I had also wanted that this fellow never existed,' replied Smith for himself. The butler did not react and continued to remain entirely impassive. He was of English origin and never mentioned "our" but always spoke about "your" subjects, such as "your" security service. He remained in his heart an Englishman.

'Put him in the library, Prince, I will see him there. The butler slipped away and Smith said to Joan: 'Peasley always brings misfortune. It is that he has served previous Presidents, otherwise I would have replaced him for a long time. I don't like him, but I cannot deny that he is competent.'

Smith went to the library. There he found Peasley in front of the window. He seemed more massive than ever and his whole attitude had something of stiffness and highhandedness. At hearing the President enter, he turned around and spoke: 'Sorry Sir to bother you, but I have a matter of great urgency. The Russian Yuri Kaspadov is busy in Washington with a zwikker detector. I've confiscated the detector and locked up Kaspadov. The Russians have gone outside their agreements and have probably installed a detector in a satellite to trace operating zwikkers. They have sent Kaspadov to Washington with a portable one to spy on us. Because my computer checks automatically all incoming people who are related with zwikkers I traced Kaspadov.' On the fat glowing face nothing could be read.

'This man holds back something,' thought Smith. 'And?' he asked.

'Kaspadov came on 21 December in Washington. The next day we observed that he had a zwikker detector in his hotel room. This behaviour does not fall under the agreements with the UN. The UN hasn't informed us either, which would have been normal. The Russians are busy on their own. I thought that you and the UN must be informed with what is exactly cooked up.'

'I asked a moment ago "and"? With that I meant "why" should this Kaspadov trace zwikkers in Washington? Has someone acted against my orders? What would you do, Peasley, if someone acts against your orders?'

'Dismiss him immediately, Mister President.'

'Thank you. What is your answer on operative zwikkers in Washington?'

'These zwikkers are in our building, Sir. I know that the study on zwikkers is permitted since science is so far that these can be developed everywhere. In US Security Service laboratory we have for this reason developed a zwikker and this apparatus has been picked up by Kaspadov with his detector. We are also busy with detection apparatus and anti-zwikkers. These are almost ready.'

'Why do I know nothing about this? The research on zwikkers happens to be under the umbrella of the Academy of Sciences and you don't belong to them, co you?'

Peasley who was still not offered a chair, wobbled back and forth and said: 'It's my duty and that of the Security Service to protect our country against bad influences from outside. The zwikker falls in that category. And without an own zwikker it isn't possible to develop a detection apparatus and an anti-zwikker. If soon gangsters can walk around with zwikkers in their hands, it's too crazy when we are unable to react.'

'But why are you coming just now to me with this tale? First of all you're already for a long time busy with the development of the zwikker and you didn't inform me. Secondly that arrest of Kaspadov happened four days ago. Why do you come to me on the second Christmas day? You can only have a very compelling reason.'

'I have, Sir. My intentions to develop the zwikker within the security service are entirely patriotical. But I waited to disturb you until we had reached a result. It could otherwise bring you in difficulties. Due to Kaspadov, I realised that from the side of the Russians a response can come which might embarrass you. For this reason I thought to inform directly. With our policy to develop our own zwikker, we have acted according to the habit of the security service. We have done the same in other cases and safeguard the US and the President many times of problems. The action of Kaspadov falls, however, under espionage by foreign power and I was obliged to act as I've done.'

President Smith could not help to admire the head of US Security Service. His explanation seemed convincing and Smith always appreciated when people showed own initiative which saved problems to him.

'Speaking about spying. What about that case in Moscow with your agent Spone? A burgling has been committed and although no one has made a report that something was missing, I haven't heard yet a satisfactory explanation what has really happened.'

'I've reported you about Spone, Sir,' answered Peasley on a tone which pronounced doubt whether his report was read.

'Indeed, I will ask my secretary to bring it.'

'I've one with me, Sir. I thought that it might come up for discussion if you dine with Mrs Guerrero. This is the main reason for which I disturb you today, so you could inform Mrs Guerrero of Kaspadov's espionage when she is dining with you tonight.'

'That you know too?'

'My job, Sir. I have to follow all your movements and foresee the consequences of it.'

Whether President Smith could read decency from his face is doubtful. He frowned at least. 'Well, well, I appreciate your report, although I have the feeling that you should have acted differently. I will discuss that tomorrow. Be here in the White House at eleven hours A.M. sharp.'

'I will be there, Sir.'

Smith rang a guard, who led Peasley to his large magmobile. The President realised only then that entirely against his habit he had let his guest stand all he time. Smith shook his head and returned to Joan who was still knitting her sock.

'And, what had our piglet for you?' asked Joan, 'certainly something sneaky?'

'You say it. He has arrested Kaspadov because he looked with a detector for zwikkers in Washington.'

'Is there then a prohibition to have detectors which can trace zwikkers?' Like always Joan came straight to the point.

'No, not that I know, but Peasley brought this under espionage. Finally they have held Spone in custody in Moscow. I don't know for certain whether he has spoken the complete truth. For this reason I have summoned him to come tomorrow morning in the White House. I think that our diner with Dolores may produce some extra information. She has been rather well with Professor Bolotnikov and it's possible they know more. Finally she asked for a meeting and we have converted that into a sociable diner at our home.'

President Smith was right. Dolores had been approached on 22 December by the Russian President concerning the case Kaspadov and she had consulted the "persons of the first hour", as she called them. President Smith was the last. She had been rather agitated about what she had heard and she hoped that she could find a solution by means of a personal conversation with Smith.

The "representatives of the first hour", Frigger, Lin, N'Goba and Bruanetti of Australia, Asia, Africa and South-America had pressed her to investigate this thoroughly. Julien De Beaufort had in the beginning reacted evasively, but came then with the disconcerting story that he possessed in his safe the stolen construction drawings of the zwikker. Dolores had been very upset and had requested Julien to send her a complete report, including the names of the persons involved. What had reassured her slightly was that Julien hadn't spoken with anybody, although the reason was rather fear for his position then a proof of statesmanship. The sealed report had been sent her by special courier. It gave an overview of the role of the US Security Service. She was anxious to learn from President Smith how much damage had been caused already. She had hesitated to give the report to John Smith. It would be sufficient to tell him that his security service had stolen the construction drawings of the zwikker.

Precisely at seven p.m., Dolores and Jorge were announced by the butler. The meeting was very cordial. Dolores was not always in agreement with the US politics, but she respected the honesty with which Smith approached the problems.

'And..., how are you,' asked John when they had gone to table. Dolores sat right of him at the round table and Joan left. Jorge sat in between Dolores and Joan. Joan noted the typical way his spouse started always a conversation. The word "and" expressed everything.

'You have been so kind to invite us upon my request to meet you. We appreciate this particularly. I have something on my liver and I wonder whether I must start with the cheese and lettuce or afterwards?' It was commonly known that the diners of the President started with cheese and lettuce.

'Why not during the cheese and lettuce. I assume Jorge can be trusted entirely?'

Dolores laughed. 'If it concerns my professional mysteries, absolutely. For the rest I will not answer for him. A writer must obtain from somewhere his inspiration?'

The entrée was served and when the butler had withdrawn, John gave Dolores a questioning glance.

She understood the hint and spoke: 'Victor rung me lately and told me that your mister muscle has arrested Kaspadov on charge of espionage. This probably in response of the arrest of an agent of his security service in Moscow earlier this year, but that's not all. It is related to the fact that you have constructed zwikkers. Victor admits that he had launched, outside my knowledge, a satellite with a detection apparatus which can trace zwikkers at distance. This satellite apparatus has detected zwikkers at three places: New York, Moscow and Washington. The last place was suspicious and he has sent Kaspadov and an assistant with two portable detectors to trace the so-called "wild" zwikker in Washington. Since I received no information from you on the legal US zwikker research, it had to be therefore an illegal zwikker. However, before Kaspadov could trace the zwikker, he was arrested. But his assistant a certain Svetlana Shornakova had. She was in another hotel than Kaspadov. She informed immediately Victor and he had this findings confirmed by a third person. Now is Victor not the person to cause immediately an international conflict. He has reflected how this could be solved. He himself apologized that he also has kept back information. But the keeping back information by the US is at least as dreadful, if not more. I assume that all kinds of declarations are possible. To make this plausible for the outside world is, however, another matter.'

Dolores remained silent for a moment. John looked uneasy. He felt himself heavily taken, although he had sensed that everything was not as it should be.

'What I propose is that both of you, I mean you and Victor make a declaration that implies that the national US research has succeeded in building zwikkers and that Russia has done a test with a detector build in a satellite. When scientists in other countries succeed to build zwikkers too, the fences are down. And mention in this declaration that each operating zwikker can be immediately traced, so one should submit oneself earlier to the UN policy than building zwikkers by themselves. My worry is whether we are already far enough with the construction of anti-zwikkers, so that "unwanted" zwikkers can be destroyed.'

'How far is Moscow with anti-zwikkers?' asked John. 'We have almost succeeded to build one, I heard this accidentally this afternoon.'

'Why this afternoon?'

'You understand that I as President will never be disloyal to my country, but before you arrived, Peasley came to me and told me that US Security Service had developed zwikkers on their own. They are operational. It were those zwikkers which have been traced by the Russians. Now you understand why I wasn't informed earlier. The research under the guidance of the Academy of Sciences is not yet so far. They keep me informed regularly.'

'And Peasley kept you aside? Certainly to protect the President?' interrupted Dolores.

'Something like that. I shall handle that tomorrow, you can be sure. But I cannot hang out all our dirty wash. Your proposal is particularly welcome to me. The zwikker can become a much more dangerous apparatus than we thought or wanted to think. I will ascertain that all zwikkers and related equipment in the US come under control of the UN and that we free funds for setting up an international detection- and destruction system of "wild" zwikkers. Only then the world can be protected of abuse. This sounds of course hypocritical if you are just rightly accused of illegal zwikker activities. But I will not run away from my responsibility.'

'In fact Dolores is very sport with us,' said Joan, 'she has bewared you for a failure. Dolores, why did you that?'

'Because she respects John,' interrupted Jorge. 'Dolores gladly acts on the basis of his intuition, her built-in zwikker, I say the last time. Happily she does only indoctrinate me, but I have bought myself a built-in detector, therefore I'm warned in advance.'

They laughed and the tension was broken. The lettuce and the cheese had disappeared without they had tasted it.

Joan rang the butler and ordered the next course. The atmosphere had become much better. The exquisite fish, which was served, could carry everybody's approval. A heavy load had fallen from Dolores's shoulders.

After his visit to the President, Peasley hurried immediately to his office. He boomed into his office to check whether the zwikker, the instruction book and the construction plans were still in his private safe. Ron Sheily had reluctantly delivered the apparatus, but he had not sabotaged it. Peasley caught the zwikker, the notebook and the drawings in a large suitcase and left his office. In the corridor with all the portraits of his predecessors, he halted before one and muttered: 'You should have done the same, do you?' The gaze of the portrait continued to stare at Peasley impassively. He got an uncomfortable feeling but recovered rapidly. Looking around to the other portraits, he went quickly to the exit. The gate keeper jumped backwards, he had not expected his boss that rapidly. Staring with open mouth at Peasley he said: 'Everything O.K., boss?' Peasley only muttered and ran outside, the suitcase holding tenaciously. He had the feeling that the zwikker made him transparent. He stepped in his magmobile and ordered the driver to drive him home.

There he stored the suitcase with its contents on the attic. The instruction notebook he took along and started to read it attentively. At two hours in the night he put it aside and went to bed. In his dream he was pursued by the Russians and all Americans laughed at him. He awoke at four o'clock, sweating over his body. There was nobody, everything was quiet. He stood up and went to the attic to look whether the zwikker was still there. It was still there at the same place behind the heap of books. A bit recovered he made himself a strong cup of coffee and started his computer. Perhaps he could obtain some information from Internet which could be useful when working with the zwikker. Provisionally he had still no idea what he should do, but one thing was certain for him, there would come a time they would need him.

He did not worry about the appointment with the President later in the morning. What could Smith do? Smith didn't know anything of the robbery. All had been done out of patriotism! Only he had to find a solution for Kaspadov. Releasing him without explanation was perhaps the simplest way. The interrogation of Kaspadov had produced nothing else then that he had to trace some zwikker signal with a detection apparatus in Washington or elsewhere. They could better release him to avoid further difficulties. Finally Kaspadov had done nothing illegal.

Peasley realised that he had acted too fast. But together with the President he would solve it. Finally the Russians did also something outside the UN. He turned all arguments four times around, but did not reach better conclusions. He faced the conversation with the President with faith. In the chair he fell again asleep and jumped up when he awoke again at nine o'clock. He had still two hours before he was expected.

In spite of the cold and snow Peasley succeeded to pass by his office in order to see whether there were further bulletins. Five minutes before eleven he drove through the gate of the White House. He was waited for by a guard and was conducted to the guard room. There he was left alone. No other visitors were there. The large clock, a present of a visiting statesman, tapped loudly. After ten minutes, the tapping started to affect his nerves and after a half hour he trembled entirely. The fat man had found his weak point and sweat stood in his neck.

Ten minutes past half twelve, the secretary Peter Scott came to fetch him. He did not excuse himself and showed Peasley in without comment at the President. These stood behind his desk and said: 'Tim,' calling him with his first name, 'sit down there. What I have to say is only between us both. I've been informed that you've acted against my explicit instructions of "no espionage" to steal the zwikker plans. A person, someone of the European Union, had succeeded to do this and he has delivered the plans in October in your hands. The setup with Spone was only to defer the attention. Remarkable enough it were not the Russians, but the West Europeans who found out.'

He observed Peasley who tried to give himself in the low fauteuil an attitude. 'Do you still remember what you answered me yesterday when I asked you what you should do with someone who acted against your orders? "Dismiss him immediately' was you answer. Tim, you have herewith condemned yourself and I dismiss you from your position. You will get a golden handshake in proportionality with your state of service and the many good things which you have done for the US. But my decision is definite, you're fired.'

What President Smith saw, was a trembling man was who only control ed himself with the largest effort. He resembled a bulldog that would attack. But whether it came because he could not rise rapidly from his low fauteuil or that he realised in time that he was with the President of the US, Smith could not discover.

Anyway he stood up with difficultly and stuttered hoarsely: 'But Mister President, I've only acted on the basis of patriotism? In my career I've never acted otherwise and always to large satisfaction of you and your predecessors.'

'Would you have accounted such extenuating circumstances for your subordinate, who acted against your orders? Yesterday you were very clear, isn't it? I'm sorry man, my decision is irrevocable. My secretary will hand you an envelope with a golden handshake and he will arrange that you personal properties from your office are delivered at your home.'

Bu... but.., Mister President...,' stuttered Peasley.

However, Smith had pressed the intercom and his secretary entered to lead Peasley to the corridor. He still wanted to give the President a hand, but Smith had turned around to the window with his back to him.

As a broken man Peasley left the White House. He felt no anger, only emptiness. His work was everything for him. And someone had taken that away from him...! Taken away, because some clodhopper in Europe had found out that he had ordered to steal the construction plans of the zwikker. Would Travenkov have done that? I could have arranged an accident to that spectacle fellow,' he growled in himself. 'But that has no sense anymore. I've now only myself.'

At home he opened the envelope and saw that the amount was ten years salary. 'Not bad,' he thought, 'in particular with a zwikker in my pocket.' This idea cheered him up somewhat. 'I'm not yet finished. Finally I have ten years time and nobody can take my zwikker from me. Or would they? Sheily and perhaps the complete zwikker laboratory might learn that quickly. In my safe they will find nothing.'

Peasley paced up and down, and back and forth. It had gone all so rapidly. It was the very last thing he had expected. 'Fired...! I can't take even my own things...! Those will be brought...'

In his complete life he had only risen. Wasn't he nevertheless almost the most powerful man of the US? And now? He tried to arrange his ideas and evaluate what could happen. If he handed in the zwikker, he would have a free life. He could go and come where he wanted. He would not be astonished if next week someone would stay on the pavement to ask for his services. But if he kept the zwikker, one could decide to fetch it back with the strong arm. To deny that he had taken it along, would hardly help. But when he made a copy and returned the original apparatus? But that would cost certainly two months time.

With this idea the colour returned on his face. The trembling of his hands stopped and he sweated no longer. 'That's it. I will disappear for two months with my zwikker and let somewhere in the world make a copy. Or I make it part ally by myself.'

He went to his computer and programmed a list of electronic specialists in South America. In Brazil there were a number which were also specialized were in metallurgy. Those he should contact. And then someone who wasn't too scrupulous. He printed the list and decided to leave the same day. In order to lead the Security Service astray, he rang his ex-secretary Harry Snowden at home. Snowdon looked baffled at his former boss.

'Can I be still of your service, Sir?'

So he was already informed by his dismissal. 'Yes, Harry, have they assembled already my personal things?'

'Not yet, Sir, it will not happen earlier than the day after tomorrow. Are you then at home?'

'That's exactly why I call you. I'm going on holiday for about two months. My intention is to travel to Ric de Janeiro and then descend down in Brazil. Perhaps I will finish in Chile. I am provisionally inaccessible. These things can be delivered to my house keeper. She's at home. They must, however, ring in advance. Harry, I thank for the work you have done. Perhaps see you later.'

Before Harry could have answered the connection was broken. Harry sat with open mouth. That was the first time in ten years that he had got thanks from his boss.

Peasley got his trunk which stood generally ready, took some credit cards, the cheque with the golden handshake and the suitcase with the zwikker. A few minutes later he left his house. He told the same tale concerning Brazil to his house keeper and he had given her two months salary in advance. She should not worry when he did not ring for some time.

A moment he considered to write to the President that he had a zwikker, and that the US could always count on him in case all zwikkers of the world had been destroyed, but he abandoned this idea rapidly.

He went with a taxi to the airport, collected his cheque into ready money at the exchange bank and bought a ticket for Rio. An hour later he was in the air.

Thus the people who brought Peasley's personal belongings heard from the house keeper that he had gone with holidays to Brazil and would return in two months. Furthermore she knew nothing. On her questions what had happened, she got no answer.

In the meantime President Smith had called the Minister of Interior and informed him on the dismissal of Peasley. The security service fell under this ministry and the Minister reacted slightly piqued that he was not consulted in advance. He was a man who hated out of experience the easy American firing policy.

'Can you rapidly propose a successor George, the position can't be left open. As soon as you have a candidate I want to speak with him. I had to settle the problem with Peasley immediately to prevent undesirable actions. We must be extremely careful with those zwikkers. Could you check whether Peasley didn't take something with him?'

'Sure, but haven't you reacted too fast. Peasley may have acted against your order, I agree, but you imposed on him a "no spying", while you know that each clever schoolboy can do that with an advanced computer. And Peasley, although I don't like him either, has served the US always perfectly. He almost thought he was the US himself. Anyway I'll try to find you on Monday one or two candidates.'

'George, I'm still the President who has to be respected, and Peasley failed to do that. You and I must together prepare a declaration for the House or Representatives and the Senate concerning the replacement of Peasley. We will not mention the robbery of the zwikker plans. The fact that he developed zwikkers on his own is already bad enough. I will now ring Professor Bolotnikov for our common proposal that is expected by the Secretary-General of the UN.

'You want to speak with me?' asked Professor Bolotnikov, giving Smith the possibility of stipulating the subject. He had been informed by Dolores that Smith would ring him up and that they should prepare a common public declaration.

'Yes, I've understood we both have been naughty. We, even more than you. On this fact I was only informed yesterday. We have developed illegal zwikkers and you have launched a satellite to trace zwikkers. I can say that I'm grateful for that. Otherwise I would not have known. Fortunately Julien De Beaufort has contributed too, therefore we can make a virtue of necessity by preparing together a declaration. Do you agree?'

'John, you've scared me. After we caught Spone and after the robbery had been detected, we were uncertain where those zwikker plans had gone. It could be anywhere in the world. For this reason it was necessary build a detection system in a satellite. We have indeed worked on our own, because we feared a delay by means of legal consultation. Commissions, etc...., you know what I mean. In fact it was lucky when we found the "wild" the zwikker with you. What did you do with Peasley?'

'I fired him. I had warned him in September for not committing "espionage". I cannot tolerate that someone ignores a command of the President, in spite his capacities. But what concerns our epistle, you agree?'

'Prepare something beautiful about the necessity of a worldwide monitoring of zwikkers and faster developing anti-zwikkers. I will answer by returning e-mail so that we can have a good document by tomorrow. We must, however, lay the emphasis on the fact that Russia and the US will soon free a larger budget. Within two months there will be launched twenty satellite detectors in space and Interpol must be equipped with anti-zwikkers.'

'In the meantime you have already dictated the complete text,' Smith laughed.

'Still one thing John, let Kaspadov go immediately and with sufficient excuses. That fellow has deserved better than being locked up by a certain Peasley. And keep an eye on Peasley, he can't be trusted anymore.'

'You can count on me. I will speak personally with Kaspadov. You have no secrecies, I assume?'

'Not where it concerns zwikkers. Well, I wait your concept and thank for your phone call.'

The first thing Smith did was ordering Harry Snowden to get Kaspadov on the phone. Obviously, it took some time to arrange that. Kaspadov looked pale and confused. Locking him up hadn't done him well.

'Mister Kaspadov, with Smith. I offer you my apologies for the inconvenience which you have experienced. This should not have happened. I have said the same to President Bolotnikov. If you want you can ring him on the spot. But we understand that you would prefer to go to your hotel. The detector, which was confiscated, will be returned to you. Once again my apologies. Later you will learn that your visit to Washington had also a positive side. It will speed up the mastering of the zwikker. But this you can learn from your President.'

Kaspadov was still confused. He had been completely surprised by the personal call from the President of the US. Part of the angry ideas which he nourished against the US and the security service disappeared. If only he was free. He had enough of this building.

'I appreciate your call, Sir. But don't take me evil when I would like to leave here as quickly as possible.' He broke off the connection.

Without further comment someone brought him to the hotel where his room was still on his name. He put down the trunk with the sensor and waited to be alone for catching the telephone. Would Svetlana still be in her hotel? Or had she been called back to Moscow? He trembled when he tapped the number of hotel President and waited for connection. A young lady of the telephone service put him through and let the number on Svetlana's room ringing for a long time. After ten times she returned in picture with the communication that nobody was present.

'Is Miss Shornakova still registered?'

'Sure Sir, and I will put you through with the counter to look if she has left behind a message.'

She disappeared from the screen and a moment later appeared someone in uniform who said: 'Here is Jack from the counter, can I help you?'

'Has Miss Shornakova left behind a message for me?'

'What's your name Sir?'

'Kaspadov, Yuri Kaspadov.'

'And you said, Shornakova, first name Svetlana? Yes, I have here a note she left behind just an hour ago. The tenth message. Each time she replaced the old one for a new one when she left the hotel. But I cannot read it, since it's inside of an envelope.'

'But Mr Jack, let me legitimise myself so you can see that I am the addressee. Please look!' Kaspadov's nerves broke almost down when he placed his card for the camera of his telephone. 'Look Mister Jack, Svetlana will be grateful if you show me the contents of her message on the screen.'

'You ask me something. The fact is, that I know that sympathetic lady meanwhile well and because she is worried about you. Here is it, I will not read it myself and I tear it directly in pieces after you have read it.'

Yuri read: *"Yuri, tonight I'm in room 555 in hotel Astoria, Svetlana."* She was just close by. 'Thank you very much Mister Jack, thanks a lot.'

He saw Jack tearing the note and Yuri cut the connection. Room 555 was just above. He stormed from his room, took the elevator to the next floor and knocked on number 555.

The door opened and an older lady stood before him. Not Svetlana.

'Is Svetlana Shornakova with you?'

'Yuri..., Yuri....!' he heard a call from the room and Svetlana blew along the lady to him. She flew her arms around his neck and gasped: 'Yuri, I was so worried about you. Oh, how delightful it is that you are here!'

Yuri held her firmly to his heart. He could not bring out word. That Svetlana would approach him this way he would never dared to dream. He kissed her and caressed her hair.

Beside him he heard a voice saying: 'Eh.., when you youngsters are ready, would you perhaps come in?'

Yuri had entirely forgotten the lady and stammered: 'Of course..., but I missed Svetlana so much.'

'It was only five days, no longer,' he heard saying a man's voice.

The lady took them by the hand and said: 'This is my husband Jenke Holthus and I am his spouse, Astrid. Cordially welcome Yuri, thus we know you from the tales of Svetlana.'

'But.., but.., Svetlana what are you doing here and just above my room with these people?'

'Yuri, sniff.., it's a long story and I will tell you everything. It's so fine to see you again. I didn't know you meant so much for me. I haven't overwhelmed you too much?'

'Oh no, I like it after all those days been locked up.'

'Shall I tell the story, youngsters or is Svetlana going with you Yuri?' spoke Astrid.

'Svetlana will tell me when we go to my room, if you allow us. We can see each other later again, isn't it?'

'You seem to be lovers. I suspected so much. Jenke and I wish you a lot of luck, isn't it Jenke?'

'Yes.., what? Yes naturally, much luck,' repeated Jenke, which had looked out of the window. He expected Ron and Ben and he was curious to know the latest news.

Ron and Ben came half an hour later Jenke opened the door. Ron's goatee hopped of emotion and Ben swung as a hypnotised snake.

'What's the matter, boys?' asked Astrid.' You seem very excited.'

Ron said. 'I was called at home by our Minister personally and we had to come to the Ministry.'

'Peasley has been fired, he could not even return to get his belongings,' interrupted Ben. 'Ron had to collect these. But the zwikker which Ron gave him was no longer there!'

'I've looked everywhere, but found nothing. He has taken it along with the instruction notebook and the construction drawings.'

'My lord, and where is Peasley now?' asked Jenke.

'In Rio, at least if we must believe that. But his name is not on any list of outgoing flights. When we rang this afternoon, his house keeper said that he had left for Brazil. She only knew that he would stay away two months. With holidays, she said. That was correct with what he said to his secretary.' Ron nodded as if he wanted to confirm this.

'What has happened, will change our plans entirely. Our complete zwikker department will be transferred to the New York State University. Order of the Minister. Furthermore Ben and I have been appointed as expert inspectors of the complete American zwikker research programme. There will be an acceleration of the production of detection equipment, both for satellites and portable models. Furthermore anti-zwikkers have to be produced as fast as possible.

'And what should we do with the zwikker you have at home?' asked Jenke.

'We can do nothing else than leave that there. It's not possible to smuggle the zwikker back, also not in components. When the inventory is made it will be clear that one zwikker is lacking. I can explain and others can confirm this that I brought this one to Peasley's room on his request.'

'So you two become an inspection team for the total US zwikker research. That's good news. Then we haven't to do things on our own. Your own zwikker must be stored well away so nobody can touch it. To assure that you won't use it, you can give some essential components to me. We can then always testify that you have acted honestly because of concern about Peasley's behaviour. I'm nevertheless delighted that my action in Moscow has contributed to something positive.'

'Now you have become colleagues of Kaspadov,' noticed Astrid. 'But what about the zwikker which has been taken by Peasley?'

'We have to consider that', answered Ron. 'We can only find that zwikker if he uses it. Moreover this man has more experience than anybody else.'

'Then it can last months before his zwikker is found. That one Russian satellite sensor has too little capacity, I understood. Perhaps Kaspadov can inform us what to do,' said Ben.

'Oh.., we haven't told you yet,' called Astrid, 'what stupid of us. Just before you both came in he was with us. He continued to kiss Svetlana.'

'Which Svetlana?' asked Ron.

'Oh, we haven't told you either,' said Astrid. 'You remember you found an indication in Kaspadov's room on hotel President. I've been passing by. It was indeed a girl. She is an assistant of Kaspadov and had also a detector. She informed Russia and thus the ball started to roll. She's called Svetlana and they came here just before you. She had left a message for Kaspadov in her hotel. Kaspadov came storming in after his release and then he started to kiss her. Their first kiss, very romantic. I find our adventure in Washington terribly exciting.'

'But where is Kaspadov now, so that we can speak to him?'

'In his room naturally. He and Svetlana had to tell each other so much, nice isn't it?'

'Astrid likes romantic people, boys,' said Jenke. 'I will ring room 455 and ask if he can come.'

Kaspadov appeared on the screen of the telephone His hair was rumpled. 'What do you want Mister Holthus?'

'We have here two agitated US security agents whom would gladly speak to you.'

'And I don't want to see any of those fellows!'

'But hasn't Svetlana explained to you who our two friends are?'

'Not yet, but she might do if you grant us some more time. Over an hour, perhaps?'

'Well O.K., but please not later,' answered Holthus. Kaspadov had already broken the connection.

Yuri and Svetlana had been both completely surprised about their feelings and for the first time in their live they felt themselves floating in the air. Just after the phone call of Holthus they returned to the ground and remembered their tasks and the confusion with Holthus. Svetlana told Yuri what had happened since his arrest, how Professor Bolotnikov had reacted, that a certain Mario had arrived to pick up the detector and that she had been spotted by Astrid. She gave an overview of what had happened at the US Security Service such as was told by Sheily to the Holthusen. Furthermore she had got a month free from Professor Bolotnikov and that she could enjoy a holiday.

'Don't you think that I must first check with Professor Bolotnikov?'

'Why? Can't we do that tomorrow or next week?' begged Svetlana, 'you will risk he calls you back immediately. And then we would be separated?'

'Or you join me and I will invent an excuse that we can work together.' He feared too he would be called back. 'But I can't do otherwise. I must ring him first.'

Yuri tapped the number which linked him directly with Professor Bolotnikov, wherever he would be. He proved to be at home and a sleepy voice said: 'Hello.' The screen stayed dark.

'Professor, this is Kaspadov. I have been released and call you from my hotel room. Svetlana is with me.'

'Yes, I see that. But why haven't you called earlier? You've been released for hours and now I fell asleep. Is everything well with you?'

'Yes Professor, do you want me to give a full report or can it wait until tomorrow?'

'Tomorrow my boy, I expect you as soon as possible in my office. Ring tomorrow at a more Christian time, say ten o'clock Moscow time, so we can make further agreements. Reserve your return trip and take Svetlana along.' He broke of the connection.

'You hear that, Svetlana, we have to stay together. Let us go up now and learn what your troika has to tell us'. Yuri embraced Svetlana and took her by her hand.

'Ben Corward and Ron Sheily,' presented Holthus them, 'and this is Yuri Kaspadov the inventor of the zwikker and his colleague Svetlana Shornakova. Sit down youngsters. We've something that will interest you.' Ron looked curiously at Kaspadov.

Ben and Ron told the last events. About the dismissal and disappearance of Peasley and about the measures which the UN in association with the US and Russia intended to take. They mentioned also that Peasley had probably taken along a zwikker. What they didn't tell was that Ron had a zwikker at home and that some essential components were given to the members of the troika.

Kaspadov listened patiently. 'I understand that your story has a certain confidential character,' he started. 'But what you said about Peasley, I will tell to Professor Bolotnikov so that he can discuss it at international level.'

'Indeed, and it would be useful when your satellite sensor is focused on South America.'

'That will not be easy with the existing satellite, but we can shortly launch a second.'

'But how can we trace Peasley? We can ask the US Security Service and Interpol to make up a team in which you both are involved?'

'And we then?' exclaimed Svetlana and Astrid at the same time.

'Peasley knows your faces,' said Astrid, 'he will smell you miles away. It is also possible that he knows my darling, Jenke. As a matter of fact Jenke must return to his work in Strasbourg. Only Svetlana and I are unsuspected persons. Therefore I propose myself for this particular espionage work. Who can hire me for much money? You in Moscow, Yuri, or you in Washington?'

'And me?' interrupted Jenke.

'You can coordinate, the point to whom we can send unsuspected messages. Nobody will suspect you and you can inform everyone else.' Astrid was entirely enthusiastic. At last something very exciting to which she had hankered all her life.

'That proposition isn't bad at all,' said Kaspadov. 'I will present this to Professor Bolotnikov. We pay perhaps better than those fellows from the US Security Service. The fact is that Henke and I have to miss you is for a good cause.' He looked at Svetlana slightly painful. Perhaps it won't be as bad as that.

But it did. When he rang about midnight Professor Bolotnikov, which was ten o'clock in the morning in Moscow, and launched the proposal, the response was positive. The fact that Peasley had stolen a zwikker shocked Bolotnikov seriously and he found that immediately action should be undertaken. Peasley could avoid any security service if he wanted. Additional unconventional aid was required. He requested once again Kaspadov and Svetlana to return as soon as possible and promised to do the preparatory work. He said to Kaspadov that it was better if Astrid Holthus returned with her spouse to Strasbourg and that the contacts would run by means of Svetlana.

A week after Christmas the common declaration of the Presidents of the US and Russia was made public. The world press reacted concerned and mentioned panic in the zwikker world. Questions were asked to governments and several countries had emergency debates.

Most concerned was the religious world. So far the reactions on the reports of the Alberta-group, as it was called, had been rather lukewarm. Some sects spoke of diabolic developments to erase God. Others required the zwikkers for personal use, so they could confirm the truth of their faith. There were also groups who claimed the zwikker for study into the soul and to be able to contact their ancestors.

China wished to use the zwikker as national instrument to cut back the birth rate to half. A State in Africa which had a fight with his neighbour, wanted zwikkers for its defence by lowering the weight of their army vehicles so they could drive through marshes. A group of arms trafficking wanted to take the zwikker in production for sale on the black market.

The result was that the reports of the research laboratories of New York and Moscow received more interest. Within a month their press editions had acquired up to ten millions copies in English, French, Spanish, Arab and Chinese. Also the demand for older numbers had increased. What hadn't been mentioned in any of the reports was the intention to develop zwikkers as an instrument against terrorism of fanatic groups.

Niterói

These and other bulletins were read by Peasley in Niterói, a city on the bay of Rio the Janeiro. He was having a beer on a terrace. He frowned by reading the English Rio Tribune. The sun was setting behind Rio and gave a red glow over the bread mountain. On the beach, Brazilian children played football. It was warm, thirty six degrees. Peasley sweated. Since his departure he had followed most of the press bulletins and the news by means of Internet. The name Peasley had not been mentioned by now. For certainty he had adopted another name, one which he hadn't used before as an agent of the US Security Service. He had had always five complete sets of documents to his disposal, so that he could change identity if necessary. In Rio he had landed as Benjamin Troover and nobody had asked questions.

He had visited already three electronic engineers with knowledge on metallurgy but none of them pleased him and he was now looking for the fourth. The address was in a house, one block behind the boulevard. He paid his beer and passed a street filled with people who went to the beach or returned. Some were in swimming suit and had a towel on their shoulder; others had been dressed in short trousers and a T shirt. Tropical fruit and coconut milk was sold at many street shops.

Arriving at the address he pressed the bell. He was let in by a doorkeeper who followed a soccer game on the threetel. At mentioning the name, the man indicated with his thumb to the elevator and said: 'Fourth floor, number four hundred twenty two.' He had spoken broken English, but sufficient to be understood.

Without saying *obrigado* he ran to the old elevator, found the number at the fourth floor and rang. The door was opened by an enormous negro who looked down at Peasley. 'Come in,' said the man with a surprising high falsetto voice in reasonably good English. 'With what can I help you?'

'Can you help me to build an apparatus of which I have an example? This means together with me? A normal knowledge of electronics is required, but I need a specialised knowledge of metallurgy.'

'Sit down Sir. My name is Antonio Pereira. A descendant of the famous Pereira's, you know I suppose. What's your name?'

'Benjamin Troover.' The name Pereira said nothing him. 'I asked you if you can help me?'

'Yes, certainly. So Benjamin is your name. You can say Anton to me. I can everything what you ask. I've never disappointed anyone, also not my girlfriends, ha.., ha..., ha...!'

'Can you be trusted when it concerns a secret task?' asked Peasley, who started to like the man.

'Can I be trusted'? Even Joseph from the bible would trust me. My customers have never complained. I don't ask where they come from, although I knew directly you aren't a Brazilian. My knowledge I have picked up at the best schools in electronics and metallurgy abroad. What I construct will be immediately forgotten when it leaves the door. Therefore come up, what do you want to be copied?'

'A zwikker, nothing else then a zwikker. I don't tell you why, but I offer you ten thousand dollars for building together with me a zwikker.'

'Have you such a thing with you? Lord, that's nice, I would gladly see such a thing and how it works.'

'How it works is none of your business. Your only task is to make the same components so that I can put them together. Furthermore I don't want any questions, understood.'

'Well, well..., no questions, for ten thousand dollars I never ask questions! So come up with your things. When do we start?'

'Tomorrow, you will get five thousand dollars tomorrow and the rest at the end. For buying components you receive another thousand dollars, more is not necessary.'

On the face of Pereira nothing else could be read then the satisfaction of someone who made a good deal. 'Splendid man, come at nine o'clock with your items so we can start immediately. He tapped Peasley on the shoulder with his large hands and asked. 'Where do you live, Benjamin?'

'Also that's none of your business, the lesser you know the better it is for you, understood! And hold your tongue, also to your girlfriends!'

Outside again, the heat struck him as a hot blanket. Peasley's shirt stuck on his back and his trousers scoured along his thighs. He ran rapidly to his hotel Ikaraí Praia and ordered a ciner on his room. He lighted the threetel for the news. It concerned climate changes in Norway. Satisfied with his results he checked whether the zwikker was still on the cupboard. He started to reflect what he would do when the zwikker was ready and where to go next.

14

Is death a mystery? Also if the consciousness of the body can be measured?

New York

'Ladies and Gentlemen, I welcome you cordially', started the Secretary-General of the UN. 'In your first meeting in Alberta only one woman was present. In this building the rule applies that for subjects concerning humanity, sexes have to be equally represented. This counts also for religions. I have left it to you to choose the women or the man as your counterpart and I hope that your choice will give a positive result.'

This demand by Dolores had given some delay but finally the second meeting could be arranged. The archbishop of Canterbury had simply taken along his spouse, but for the Catholic Church and the Islam the matter had been slightly complicated. Next to the Nuncio Altaldi and Cardinal Benedictus sat two nuns, who had studied theology. Brill Dowson had chosen a boyfriend with equal humanistic conceptions.

'Before you meet, I want to inform you on behalf of the UN on the developments around the zwikker. Thereby belongs a visit to the research centre in the Central Hospital, where you can observe the functioning of the zwikker with your own eyes. The scientists, under the guidance of Professor Larson, are prepared to answer your questions. What will interest you is the prerecording of a death-bed of a Professor in psychology, who called this "my last contribution to science".

'Are we obliged to attend this demonstration?' asked the Nuncio. 'Death is for us a mystery and doing experiments with dying people is against our conception.'

'It's not an experiment, isn't it?' said Krishnamurthy, the Hindustani, 'I understand that they will only show us what the zwikker has detected. I must say it interests me deeply. And, Monsignor, it cannot do any harm if it makes you wiser. You say that human knowledge stands in no proportion to the knowledge of God and since death is for you a return to God, you must therefore be interested in this information.'

'My question had a general value, Mr Krishnamurthy. Personally I've no objection, but people with consciousness objections must have the occasion to withdraw.'

Professor Larson had prepared himself very well on this visit. The video reproduction was almost lifelike. He gave a short introduction in which he made clear that the pictures would speak for themselves. Then he gave to Klemarov, the technician, a signal.

The lights in the room went out. Since the three-dimensional screen covered half the wall, it seemed whether one was really looking into another room. On a bed lay a grey thin gentleman. He was nicely dressed and carried a small butterfly on his shirt He didn't look as someone who would die. He was supported by pillows: 'Dear observers, I know for certain that I will die very soon. I'm Professor Presume Mansfield, psychologist. My sickness is incurable and I've only twenty four hours to live. I'm fortunately still clear of mind and I have professionally requested to be measured by the phenomenon zwikker, which occupies so many people. I hope to be able to explain the peaks of my subconsciousness spectrum. The scientists of this hospital have been so kind to make a psychoanalysis and it gives me a great satisfaction that I can add something by dying.'

Professor Mansfield panted heavily and took some rest. 'I will ask to show my subconsciousness spectrum. Each component has been characterised with a denomination. The large one on the left is related to my youth. I was frightened in dark and was that still shortly ago. I'm grateful they have compensated this complex yesterday. It might be terribly obscure..., soon....' Professor Mansfield chuckled in the camera.

'Somewhat to the right is a spot which has to do with my religion. I'm, as it happens, a strictly orthodox Protestant and have a large faith in God. With my authorisation they have compensated this spot and later restored. And indeed, after the compensation I had lost my faith consciousness, although I remembered all details. On itself I could have lived happily with it, as far as there was still time, since I was entirely happy. But since I had put that condition of restoration, they restored my faith consciousness. I feel myself now again as orthodox as before.'

One could see that speaking tired Professor Mansfield and a nurse came along to check his pulse. Mansfield remained with closed eyes and gave a sign that the conversation became difficult.

The following images were divided in two. On the left part they saw the room with Professor Mansfield. Beside him were some flowers. There was further nobody in the room. This had happened in mutual consultation. The presence of family or nursing staff would disturb the measurements. On the right screen was a clock and medical recordings. Mansfield had not much family, only one son. This one was present in the laboratory. Mansfield was provided with a microphone, which hung above him.

Professor Larson gave now further explanation: 'Our patient is very weak and is in a situation which cannot last very long. He is mentally still O.K. and now and then he gives orally what he feels. He does not have pain, he feels himself slip away.'

On the screen the first consciousness spectrum appeared. It was the same as what had been shown before. It was also projected on the ceiling above Professor Mansfield. At these first pictures he said: 'I feel myself weaker and weaker, but I remember images of former days which I had forgotten. Those must have to do with those peaks left or the screen. What I see, and what you see, too, there in the other room, is that my total spectrum is still intact. Very interesting. I see also my son as a small little boy on my lap. He is ready for going to bed. You remember that Evan?'

'Yes, Daddy,' one heard answering a voice.

Dolores and the people of Alberta-group looked fascinated at the screen. It was deeply moving to hear someone explaining his own dying. They heard him whispering still a few words: 'Seeing grey, all around   , nice red colour as of a flame   , small figure   , relax......'

On the next images the clock on the right part of the screen had moved an hour forward. 'Professor Mansfield has slipped in coma a quarter of an hour ago and he breathes very difficult. His pulse is irregularly and weak.'

The tension in the laboratory was high. The nuns did not dare to look. One saw that in the large original spectrum some parts disappearing and returning again. This happened especially in the right part of the most recent consciousness. The left parts remained stable.

'Now comes the moment that Professor Mansfield is passing away. Please pay attention because it goes almost imperceptibly.'

Professor Mansfield lay peacefully. The subconscious spectrum seemed to vibrate. Suddenly the pulse stopped for some seconds but restarted weakly. Just as it seemed that this would become again normal, it fell suddenly quiet. Professor Mansfield had just moved his hand slightly as if he wanted to signal that all was over.

The spectrum remained remarkably intact for about ten minutes. It no longer vibrated and stood stable on the screen.

In deathly silence everybody looked at it, only the clock on the screen continued to move. It lasted almost ten minutes before one saw a gradual change in the spectrum. It was almost as something which flew slowly away without showing where it went. The clock designated one hour after death.

'Your measuring has proved that the consciousness and subconsciousness remained considerable time after death, do they?' asked Klaus Eckelhof, representative of the Lutheran church.

'The zwikker cannot lie, and I must therefore confirm this fact. But whether it has to do with a remaining brain capacity or with the spirit of Professor Mansfield, this cannot be distinguished with the zwikker. One knows that the brain rapidly dies due to lack of oxygen. The measured spectrum in the last period up to several hours after dying has therefore possibly to do with the spirit of Professor Mansfield. The fact that people often still feel the presence of a deceased could confirm this.'

'Professor Larson, we thank you very much for this emotional demonstration,' spoke Dolores Guerrero. 'I cannot thank Professor Mansfield, since he has passed away, but perhaps his spirit senses our thanks.' She had been deeply moved by the demonstration and she had said what she thought.

Professor Larson requested his guests to follow him to a guest room where drinks go were offered. The tension had been broken and the guests started to discuss freely with the scientists. One could see that most of them had a feeling of relief. Nobody had expected much of the apparatus, but the results were worth a reflection.

What interested them most was that their faith was part of a spectrum and that the zwikker could modify it.

On the question by a scientist whether they would be voluntary for an experiment, the answer was negative. They were too happy with their own faith by letting it damaged by an apparatus.

Professor Larson gave subsequently an overview of the hopeful results in the field of phobia of patients. In fact the zwikker could do everything what was possible with hypnosis and other alternative methods. Only better, because it was certain that an erased phobia could no longer return.

On the question if one could study by means of indoctrination of knowledge the answer was negative. 'Knowledge is stored differently in the brain then consciousness. Knowledge and memory do not produce a "field" that can be registered by the sensor of the zwikker. However, since the consciousness is linked to memory, one has to do additional research for further judgement.'

The subject material having a field detectable by the zwikker was also brought forward. Cardinal Benedictus asked whether the zwikker could measure something with relics and hosts. Jitah Crita, the Lamaist, wanted to know whether historical events such as a battlefield left behind a measurable field.

Krishnamurthy, the Hindustani, brought up the subject of reincarnation and although the Nuncio and the Cardinal were opposed to discuss this, a lively conversation took place. Krishnamurthy suggested to examine people, who knew they had recently been reincarnated, and compare their spectra with still living family members of the original person. Exactly like DNA research between family members.

Professor Larson noted all subjects which were mentioned. The discussion appeared to be useful, although he had the impression that certain subjects were scrupulously avoided. He wanted already to stand up and give his guests the opportunity to leave when one of the female guests, Brill Dowson called his attention. 'Would it be possible to measure God? I can imagine that my religious colleagues avoid this question or don't dare to ask, but seriously, would it be possible?'

Professor Larson saw the others startled. He looked at the list of names of his guests and behind the name of Brill Dowson stood humanist. 'You do ask me something, Madam, or is it Miss?'

'Miss, but you can say Madam if you like. There are too few men to marry. Part of them is priest, another part is homosexual, some are scared to take a responsibility for a family and the rest is already married. I bet on the second round, the divorced men like a second option.'

'Interesting conception. I will remember that. Perhaps there are statistics. Anyway you show realism. But returning to your question whether God can be measured, we cannot give an answer. At least not yet. We have, however, thought about it. But let me frame an answer. In the first place the zwikker would sense an enormous field that is omnipresent. Perhaps it detects more in churches or mosques. Thus the zwikker must measure something additional during the prayer of a person or a lot of persons. But then you can also argue that it is only the common consciousness of the people which is detected. You see, it is technically not so simple, because in a complex spectrum, a small difference can also be caused by electronic noise in the apparatus. I admit however that it implies a challenge. There is of course the danger that the outcomes are not in accordance with existing religious conceptions. On the other hand, if we measure nothing during prayers or with relics, this does not imply the absence of God. The zwikker has its restrictions too, at least technically.'

'If I understand you well, the zwikker is only able to measure God if he manifests himself in a "condensed" form. Such as during a Maria appearance. We can no more measure Moses with the burning shrub. Faith in God cannot be proved therefore. Or not yet?'

Indeed, you can put it this way, Mrs Brill. I even think that your colleagues cannot oppose this thesis,' answered Professor Larson.

'Would you authorise me another question? Why don't you investigate a medium, clairvoyants and hypnotists? They would certainly be good clients.'

'A good question. They are already on the research list. But we must distinguish in advance the charlatans from the legitimates.'

An Indian lady asked: 'What happened further with the spirit of Professor Mansfield. You have shown that there was still something of a spectrum up to several hours after entering death.'

'After seven hours it became time to take away the body of Professor Mansfield. The presence of the persons of the funeral service disturbed of course our results; therefore we have turned off the zwikker temporarily. We have afterwards continued the detection in the empty room. The zwikker gave the same result. One could say that the "spirit" of Professor Mansfield was still present. The spectrum decreased just later in intensity. Not in the breadth, only in height. Approximately fourteen hours after dying, seven hours after removing the body, the spectrum had therefore disappeared entirely and we couldn't measure anything more. For certainty we left the zwikker standby until twenty hours, but the result remained the same.'

The Archbishop of Canterbury took this occasion to end the discussion by saying: 'Professor Larson, we have very much appreciated your information and we have been deeply moved by the way in which you conduct your investigation. We have got enough substance to discuss for some days and we will gladly withdraw for a closed meeting. By all means we thank Mrs Guerrero, and he bowed to Dolores, for offering us hospitality in one of the UN building meeting rooms.'

'Are you coming, my dear,' he said to his spouse who in fact would have wanted to stay longer. She still had a number of questions.

'Well John,' she replied and gave him an arm and followed the others to the exit.

Dolores took Professor Larson aside and pressed him warmly the hand. 'You stutter no longer, Professor, have you cured yourself or were it your colleagues?'

'My colleagues, with the zwikker.'

'That's what I thought already. I suppose they destroyed that bit of spectrum so it cannot be introduced in another?'

'Sure, don't you worry. Nobody is in danger.'

Dolores smiled and followed the others.

Professor Larson thought: 'She is able to zwikker without having an apparatus. And she is beautiful, too.'

The Alberta-group met the day after in plenary session. The Presbyterian Vicar Whitewater had been chosen as chairman and he started with: 'Ladies and gentlemen, the subject "is there life after death" is the first one on our list. The demonstration of yesterday obliges us to discuss it in depth, based on the data we've got yesterday. The question is which studies we should recommend or dissuade. It is not necessary that we are unanimous in our suggestions. We have no control over the zwikker experiments, but we can announce our wishes to the Secretary-General of the UN.'

'Applies that too for experiments we are or one of us is against?' asked Ishmael Mohammed Jamil, the Egyptian and a convinced Muslim.

'Of course. Everyone of us is entirely free to express his or her opinion. We can give also a value judgement or a condemnation if you desire.'

'I've yesterday already made a proposal,' spoke Brill Dowson, 'according to me all religions have only a meaning with the existence of God. Each proof that can lead to the existence of God must be done first. Myself as Humanist has always doubted the existence of God. I consider it as an invention of people influenced by mass psychosis and indoctrination. But

I'm prepared to reconsider my opinion if one can prove the opposite.'

She read an accumulation of emotions on the faces. The faces of the Muslims and Catholics produced horror, those of the Hindustani and Buddhist expectation.

Whitewater frowned his eyebrows and said: 'Go on Mrs Dowson, it has only sense to make a proposal which is concrete and which can be executed.'

'Thank you, Mr Whitewater, indeed you are right, I must be concrete. We might consider the fact that the zwikker can only detect something when there are persons available, living or dead. After removing Professor Mansfield's body the zwikker observed still something for some time. Yesterday I said that if God exists, something should always be detectable. But this is not the case. What can we then conclude provisionally? The zwikker is not sensitive enough to measure God, or God can be only measured by means of a person.'

'Your analysis becomes interesting, please go on,' interrupted Whitewater.

'Thus the zwikker should be able to measure something additional with a prophet and I mean a living prophet, not a dead one. But since this is more your area then mine....,' and Mrs Dowson nodded to the representatives of the different religions, 'I leave it by this conclusion.'

There fell a deep silence. Everyone had understood the range of her observations.

'What will the outside world say of this,' interrupted Khalida Zerhoumi from Algeria. 'I'm deeply worried about this type of discussions. Our Muslim doctrine makes it clear that we must beware for suggestions which deviate from the Koran. And this seems to happen here.'

'And what about that?' said Young Lee Kim, the Buddhist. 'If one is afraid for the truth, then there is something wrong with the doctrine. But it can also confirm your faith. Moreover there are already many radical ideas in the press. You must type "zwikker" and "god" on your computer and there appears a list of articles which go much further. Only for that reason alone, we must pursue our discussion and reach reasonable proposals.'

Cardinal Benedictus stood up and asked: 'Dear colleagues. I have the mandate of Pope Paul VIII to discuss all facets which have to do with God. The Catholic Church is prepared to allow the zwikker in her churches during the baptism, prayers, consecrations and the communion. When other religions would take the same initiative, we would really progress.'

Nobody gave a direct answer. Authorisation would imply an affirmative or a blame, and this would have its consequences.

'Aren't you taking a great risk, dear colleague?' asked Brill Dowson. 'You assume that the host is really the body of Christ. Suppose nothing can be detected?'

'This risk we must run Mrs Dowson. Our faith will anyway be stronger than an apparatus, how ingenious it might be. But what Pope Paul VIII wants in particular, is to come closer to the truth.'

Mrs Zerhoumi gave a sign to the President and intervened: 'Mr Benedictus, permit me call you this way in this meeting. This is from my side not a lack of courtesy, but the word monsignor is not appropriate with my conceptions. I cannot be against neither for your proposal. But I want to prevent that somewhere in the world an opposition arises that would have a snow ball effect. There needs to stand up but one Muslim leader who banishes your plan as diabolic and all your Catholics in the world will be exposed to attacks. And the history has learned where fundamentalism can lead to.'

'But, Mrs Zerhoumi, we no longer live in the twentieth century!' exclaimed the Cardinal.

'When Muslims think that you act against the laws of the Koran, Mr Benedictus, it does not matter in which century you live. You live as a Cardinal in Rome far from the reality in our countries, in spite of all worldwide information. I know very well that we can be informed by Internet about all subjects. But the question is, do we want to be informed? I have to leave therefore this meeting when the Alberta-group accepts your proposal.'

The tension increased suddenly. What first was a sociable discussion was now a confrontation which could have serious impacts.

'I'm not taking back my proposal. I can't do that, what the impact might be. But if you think you should leave our meeting, then I will respect that, although we shall regret this as a group.'

'I'm not entirely with Mrs Zerhoumi,' spoke the Egyptian Jamil. 'I'm a devoted Muslim too. Science gives knowledge and our prophet Mohammed was not frightened for knowledge. What did not frighten Mohammed needs not to fear us. And moreover the studies are done with Catholics, who for us are no true believers. For many Muslims they are only unbelievers and studies with unbelievers cannot harm us.'

'I help you hoping that, dear colleague, but not all people haven't had the same training as you and me. And there lies the problem. I can subscribe your opinion, but I cannot warrant the studies, since I have to do that from my own conception. It does not matter whether it is done first with Catholics or other unbelievers. One step further and the zwikker is setup in a mosque, you understand?'

The President, Vicar Whitewater took the stage and tried to summarise the discussion. However, he could not influence the decision of Mrs Zerhoumi, and the result was that she left the meeting the same afternoon and returned to Algeria.

The mood in Alberta-group was depressed after resuming the discussions in the afternoon. One had hoped to advance. Cardinal Benedictus said that his Church would in any case proceed with the studies. Depending on the results, the Alberta-group could always decide later what to do.'

'This is particularly disappointing,' spoke Brill Dowson.

The Hindustani, Krishnamurthy concluded it differently: 'Dear colleagues, we don't let us discourage by one derogatory opinion or threat, do we? We in India will support studies on reincarnation. Although I've understood that Christians rejects reincarnation, they insist that new-born children should be baptized to clean them from a so-called original sin. But where does that original sin comes from? From a former live, or because their parents had sex, which seems to be a sin for Catholic priests? We can check that by new-born babies with the zwikker.'

'And if they got that from their mother?' noticed Brill Dowson.

'All the more reason to examine them together with their mothers, to start with pregnancy,' answered Krishnamurthy.

'Anyway,' replied Brill Dowson, 'all faith handlings haven't improved better people. Furthermore I would like to bring forward the following consideration. Suppose reincarnation is just a matter of consciousness. Parts of it are picked up by a person, which gives him or her the idea that he or she has experienced it in former days? Logically that someone thinks he is reincarnated, with the emphasis on thinks. Consciousness and memory are not entirely disconnected, such as I have read from the de-stuttering of Professor Larson.'

At this moment the Chairman interrupted: 'It is the question, Mrs Dowson, whether your suggestion on the original sin is appropriate with respect to the zwikker.'

'Sorry Sir, I have more. For you as believers, a number of conclusions are obvious, but not for me! My question is: If the human consciousness is coupled to a divine consciousness and the other way round, to which divine consciousness was coupled the world when people did not yet exist? In time dimension, mankind exists only for two millions year, while life exists since some billion years. Here something is not correct. I don't know the solution, but I would appreciate if we can consider this as a group.'

Vicar Whitewater looked at Bill Dowson slightly astonished. 'I've more often heard this idea, Mrs Dowson. But as I said before, it is the question whether this falls within our task. Your point concerning reincarnation seems to me worthwhile to be examined. For this reason I conclude that the Alberta-group proposes to the UN two specific zwikker studies: one related to consciousness in the Catholic Church and one related to reincarnation with an Indian organisation.'

After this the meeting was closed and Vicar Whitewater informed the Secretary-General of the UN.

'How was your day?' asked Jorge after the diner. 'You were with the Alberta-group, wasn't it?'

'Yes...., indeed.... but their meeting went petered out. One Algerian withdrew herself from the group.'

'Why was that?'

'She didn't agree with any studies about the presence of God. Chairman Whitewater, a very nice man whom really looks like a Vicar, came to me to submit me his report.'

'What are you going to do when indeed riots break out as a result of the studies? Have you thought about that?'

'I do. Perhaps we might be able to bring people to reason with the zwikker. But are we on the right track? Oh, Jorge that zwikker makes me so tired. And the box looks so innocent. Yesterday at the threetel video of the dying Professor Mansfield everything was very emotional. We had the feeling to walk outside the borders of our existence. But if the thing whips up groups of people, where we will end up?'

'People will always react that way, zwikker or no zwikker.'

'It's not that simple. By the way, a zwikker is missing at the US Security Service. Peasley has disappeared and with him a zwikker.'

'When has that happened?'

'The end of last month. Interpol is of course involved, but the track stopped at the airport of Washington. Peasley said he was going to Rio, but he hasn't arrived there. He's an experienced veteran and he knows how to make himself invisibly. His name hasn't appeared in any Brazilian hotel, neither elsewhere in South America. He can be everywhere, in Africa, China or nevertheless in the USA. The programme to trace zwikkers has been accelerated and soon twenty satellites will be launched in space. So far we haven't measured any activities in Brazil with the first satellites. I'm really worried. President Smith shouldn't have fired him. Peasley was in spite of all the best men to protect America. But you know, it's still a bad habit in the US to fire people. No democracy at the work floor.'

15

Women can succeed by their intuition where men will fail.

New York

'Mrs Guerrero, I have here Professor Bolotnikov on the line,' it sounded from the intercom. 'Can you take over?'

'Yes, Soraja, put him through.' She had arrived refreshed at her office. The morning had been clear with light frost on the ground. Everything indicated to a beautiful day.

'Hay Victor, how are you? Your face looks naughty.'

'That's correct. You know that Interpol is after Peasley. Until now they haven't found a singular trace of that fatty.'

'What's your plan?'

'I want to put two ladies on his track. One of them has particular capacities. She's called Astrid Holthus. The other one is Svetlana Shornakova, you remember the violinist. I will send you the details by coded email.'

'Where is this lady at present?'

'Here near me. It can, however, happen that the ladies require help during their search. They must be able to count on you and me. We must therefore inform the security services. I assume that these won't like it, but since their agents haven't reached any results, we should use each opportunity which is offered. You agree?'

'If you say so. When will they start?'

'Today, the plan is they leave for São Paulo where Svetlana gives a concert. That had been arranged some weeks ago. Astrid goes along as her assistant and hairdresser. After the concert they go to Rio and must try to pick up the track of Peasley. That's where he went, as he has told his maid.'

'That track is already stone cold. Has it been checked under which name Peasley travelled?'

'Yes, they checked a long list of names which he used in former days when he was a US Security Agent. None of those names has been used in hotel registers or on passenger lists, wherever in the world. Rio is our only starting point.'

'Well, I will inform Interpol as soon as I have received your details. They should be very careful. You can't mock a man like Peasley. How is it, lots of snow?'

'Packages, more than we are used to the last years. But that is a subject we can discuss another time.'

'Victor, something else about I'm worried. One Muslim lady has left the Alberta group. She mentioned that the zwikker is a danger for fundamentalism, which could lead again to violence and terrorism. She didn't want to take any responsibility for such a development. I think we should be very much on our guard, perhaps sooner than we think. I think it was a great mistake of Smith to fire Peasley. He, together with Yesin could have prepared a thorough plan and action. But these Americans still have the bad habit to fire people without any reconsideration.'

'I'll contact Yesin immediately, and let you know later.'

Dolores felt a feeling of relief. At least a friend who understood her.

Moscow

In his office Professor Bolotnikov spoke to Svetlana and Astrid. 'O.K. ladies, you have the green light. I will arrange with Nicolai Yesin that you acquire sufficient aid for your travel.'

'We don't need that,' spoke Astrid. 'We've enough money with us. That's more than sufficient. I've your phone number, so we can reach you always. Everything more can give suspicion. And you've said that Peasley can smell an agent at one kilometre distance. I would appreciate, however, to have his antecedents. It's good to know the habits of that man. You have that?'

'That's no problem. In 15 minutes we can give that you. You've enough time during your travel to study it. But please destroy is afterwards. That's vital.'

'Beautifu, Svetlana, it becomes exciting. I enjoy it now already. Do you know Mr Bolotnikov, or must I say Professor,' she put her hand for her mouth, 'I always wanted to be a spy.'

'You can simply say Mister. But why not calling me Victor, so I go for a friend or family member. No professional agent would call me this way. That can be an extra protection.'

'Svetlana, we can say Victor, how do you find that?

'Not so rapid, Astrid. Svetlana will continue to say Professor. She is a famous violinist of Russia and I am for her the President and we must keep it that way.' He wondered whether this impulsive lady could reach a result. Wasn't she too enthusiastic? But he didn't have much choice. She had judiciously reacted by rejecting all aid in advance.

'Ladies, you both leave this afternoon, therefore if you like, you can still say farewell to your fiancé, respectively spouse.' Professor Bolotnikov stood up, shook Svetlana and Astrid the hand.

Astrid would have appreciated to talk somewhat longer, but she understood that a President has more to do than chattering with ladies. She had already taken farewell from Jenke who had stayed in Strasbourg. This was different for Svetlana who called immediately Yuri and asked him to come to the Red Square. How she would miss him. After their return from Washington they had become engaged and they had planned to marry soon. The concert in Brazil lay as a stone on her heart. However, she understood there was no other choice.

Yuri showed up after a half hour. Astrid had already gone to the hotel from where they would leave.

'Svetlana, come with me. I want to spend these hours with you. Professor Bolotnikov has giving me a ring, therefore I know already everything. He asked me to give you a pocket detector to trace zwikkers. I have it with me and you can put it in your hand bag. In a hotel you must connect it, however, to a threetel The range is no longer than two kilometres. Don't loose it. They have been made on the basis of a new principle, where the plasma part has been brought back to only one square millimetre.'

'Oh Yuri, Brazil is so far away from you. I want to stay here with you. Can't you come along?'

'Of course not, dearest. If there is someone who is on his guard, it is Peasley. What might help is that in a week we will have all twenty satellites in space. Those will support your search.'

'Can I call you, however?'

'Not directly, only by means of the number of Professor Bolotnikov. You can however, ring, your mother. I will pass by each evening so we can see each other and speak. It is hard, but it will pass by. And then we marry directly. At least if you still want me.'

'She caught his head in her hands and cuddled him. Thus they remained for a considerable time until Yuri brought her to her house to catch her trunks. Then they went to Astrid's hotel. She waited already for them in the hall. They had to hurry now. Checking in on the airport took still an hour for long flights such as their's to São Paulo.

Svetlana took farewell in tears and slid beside Astrid in the taxi magmobile. She continued to wave until the face of Yuri disappeared.

'Come on girl we will have a fine time together. You haven't forgotten your violin and your music? I have got lots of money, so nothing can happen with us. In São Paulo it is summer. And after your concert we will swim in Rio. Exciting. How Jenke will envy me. But that lovely fellow must stay with our children. Comfort yourself. Now you can still travel, soon have you a couple children hanging on your skirts. That's nice too; I had a fine time with them. But it's nevertheless sometimes boring to sit at home. Do you know that I look forward to this travel?' Astrid chattered on and her enthusiasm worked contagiously on Svetlana. Gradually her sorrow disappeared and a smile came on her face. With this travel she could pay a complete outfit.

Checking in went smoothly and after an hour they were high in the sky. The plane was one of newest and flew approximately four times the speed of sound. The passengers had the feeling to be in a luxurious hotel. There was a system of beds, which could be made in one turn from ordinary seats. Svetlana and Astrid had two seats beside each other, which folded out to two beds above each other. After diner they folded out their beds and fell asleep after all emotions of the day.

Astrid was reminded in her dreams that she had forgotten something important. She should have learned karate before she started this adventure. She woke up but considered that she could learn it also in Rio. One blow she had to learn which could knock down a gangster. Satisfied with this idea she fell again asleep and was only awoken when the speakers announced the landing.

Niterói

On the boulevard of Niterói sat again puffing Benjamin Troover, alias Peasley. There was no wind. Dark clouds hang over the bay, and these would certainly grow to a considerable thunderstorm.

'You can't trust anything here, not even the weather,' he thought. He examined his nails and saw that they were halfway gnawed. Since his puberty he hadn't done that, but the stress had caused him to fall back in that old habit. He saw now how he had always been dependent on an oiled organisation system, where for every task someone was available. Here in Niterói he had to do almost everything by himself. Even the cooperation with Antonia Pereira had been very stressing. The copy of the zwikker had become ready yesterday. He had hesitated to test the thing in the presence of Pereira. Moreover he was afraid for the detection satellites. He had read that there should be twenty in space. The chance to be discovered was large. There was only one possibility, to move after each test so that they couldn't trace him easily. He had not yet a clear plan. Perhaps it would be better to lead astray the fellows in Washington, New York and Moscow, by returning the stolen zwikker, for example.

Due to the stress he had slimmed a number of kilos. To avoid recognition he had allowed to grow his beard and his skin had become dark brown from the sun. Pereira had worked rather rapidly and to Peasley's stupefaction he had reconstructed the essential component of the zwikker, including the gyroscope in two weeks. He wondered whether Pereira could build now a zwikker for himself. Anyway not fast, because the sensor had been built by Peasley himself.

He had studied the guide intensively. The text on the recognition of spectrum components he had found most difficult. His plan was to measure first his own spectrum and subsequent measure some sect. There were enough of them in Brazil. He could compensate them with the standard behaviour of an average Brazilian.

Thinking about this did him good. Looking at his hands he should compensate his own nail-biting complex. That would be enough proof that he could handle the zwikker.

Peasley paid his consumption and returned to his hotel. The two zwikkers stood in his room on his desk. They looked very different from each other. The copy had been built in a tool box and was connected to a small threetel which he had bought. The original was a silvery box with an own display device. Each of them was connected to a minicomputer. Peasley started the original zwikker and aimed the sensor at himself. After five minutes he did the same with the copy. The spectra on both screens were completely similar. He stopped both zwikkers, but let the display devices illuminated. Thus one could catch no more zwikker signals.

He coupled both computers to each other and transmitted the spectrum of the original zwikker to the screen of the copy. Once again both spectra were completely equal. Satisfied he saved the spectra on an extra memory key. Further use of the zwikker could wait now.

He packed the toolbox with the copy carefully in a carton box. 'I'll put this away in the safe of my bank in Rio. I'll return nothing! It will anyway not be appreciated and might lead them to Brazil.'

After he had sealed the bundle, he took it along to the post office behind the hotel and did send it to his bank in Rio. Subsequently he asked the bill in the hotel, caught his luggage and left within an hour Niterói in his heavy terrain magmobile.

If he had known that someone would track him down, he would have started less relaxed. He had not given information to where he was going, but he had loaded the magmobile in front of the hotel. What he didn't know either, was that he had had luck with the satellites. Accidentally that moment no detection satellite hung above Brazil. One had not yet completed the programme of the launching of the twenty satellites.

After leaving the city he took the large highway south in the direction of Uruguay and Argentina. Beyond Rio he changed gear on automatic and almost soundproof his carriage rustled over the road. There was much traffic, but since it had the same speed it seemed quiet on the road. He looked at the fascinating Brazilian landscape, and started Internet on the board computer. Except tensions between a number of people in Africa, there was no news.

Passing São Paulo, Peasley searched for news on sects. One report interested him particularly. The sect had a meeting hundred kilometres south of São Paulo. There were complaints from the surroundings and they suspected dark practices with animals. He looked on the map where he could stay and programmed his magmobile computer. Then he arranged his chair horizontally and fell asleep.

São Paulo

'What did you play extremely well, Svetlana. Do you know it made my flesh creep. And that for contemporary music. The orchestra accompanied you terribly well, but you really stole the show. Beside me sat a lady with tears in her eyes. You're very talented.'

Astrid sat with Svetlana in her hotel room in São Paulo. The concert had been in a hundred years old concert hall, where approximately two thousand people could find a seat. The form of the hall had been based on a 17-century theatre in Parma, Italy, which had extremely good acoustic conditions. With two thousand people in the theatre the cadence, which played Svetlana without orchestra, was so pure that any electronic regulation could not have improved that.

'How did you find it yourself?'

Svetlana hesitated as if she searched for the correct words. 'It was as if I floated in space and that the consonances filled the complete building until the high wooden ceiling. I've never experienced such acoustics, there was hardly an echo. All consonances were fuller, or better perfect.'

'The rest of the concert was also beautiful. That symphony of Rogério Garavaglia was moving. Who was Garavaglia?'

'A composer from Brazil. You know, one of the new avant-garde. After the peaks in the seventeenth up to nineteenth century, concert music has had few good composers. The development of the electronic music and a lack of interest led in the 20th Century to a depth point. But at the beginning of the 21st Century there came a change. The compositions became again harmonious and music started to touch again the harts of people. The first composer of this avant-garde was a Japanese, Yu Jamodi. He had little success until his 40th year, but had a break through afterwards. Yu was able to combine the ancient laws of baroque music in a masterly manner with those of later composers and added to that his own genie. Garavaglia is one of his followers, although he has clear South American melodies in his music.

'Say, that's interesting. Since I'm now your hairdresser, I will deepen myself in your music. But to remain practical, we still have to eat? Tomorrow it is a busy day. We must start our tracking in Rio.'

Svetlana nodded, and found it pleasant that Astrid was with her. This woman seemed to have an inexhaustible energy and she was lively and sociable. She didn't look forward to go to Rio, but she wanted to show Yuri that she could act if required. Only just not tonight. The concert took nevertheless more energy than she'd expected.

Together they went to the restaurant of their hotel. It was still open in spite of the late hour. Svetlana was overjoyed of seeing all that fruit on the buffet. There was also fruit in Moscow, but not such as these, which were ripe and tasty. Astrid and Svetlana enjoyed their free late evening and continued for long discussing the concert and what they had to undertake tomorrow.

'How do you intend to trace Peasley?' asked Svetlana.

'I think we should start at the airport of Rio. Probably he has changed his name already in Washington, but he will still look like a pig. The passengers who leave and arrive are registered by video. We must check all old videos.'

'And if we see him on a video, what then?'

'Then we must find out his name from the passenger list. With some intuition we must succeed.'

'I like you optimism, but what then? Then we must search for him somewhere in Rio or elsewhere. If he has registered himself at least under that new name. If I was him I would do something about my appearance. A pig with a beard resembles no longer a pig. There are thousands of men who are bold and have beards. Sometimes I think they do that to compensate their boldness.'

'For this reason he can be twice as dangerous. Therefore we must watch out as frail women with him!'

'Do I have a particular task?'

'Not yet, later. But.... what would you say of a good night's rest. You have deserved that my child. What you have played fantastically. In fact you are much too good for spying. With your talent you can give to people something valuable which is good for their soul. If we had a zwikker we would be able to measure it. Will you continue giving concerts after your marriage?'

'Of course, I can't miss it. I hope that Yuri and I will get two children which are also musical. But you're right, let us sleep first.'

Astrid and Svetlana went to their rooms and fell deep asleep on their floating beds. Svetlana slept dreamless, but Astrid dreamed of bearded pigs which became bigger and bigger until they came floating like a balloon. In her sleep she giggled at the idea that she would prick the floater and what kind of bang that would give. After just awaking slightly she turned on the other side, thought of Jenke and knew nothing more.

The next morning she was up very early. Through the open window the smells of the city came in, tingling and slightly alcoholic. Because São Paulo lies somewhat higher, it is in summer not as warm as in Rio. The scents came from the vehicles which run on alcohol. The incomplete combustion ensured the light alcoholic scent in the atmosphere. Astrid looked out over the city with its immense skyscrapers.

She turned away from the window and took the telephone to ring Svetlana. She appeared with a sleepy face on the screen. 'Hello, who's there? Oh, Astrid, why do you awake me already?'

'Do you know how late it is? Already eight o'clock. We must soon go to Rio. Work is waiting. You hoist your bed and come over twenty minutes downstairs. I propose we leave in an hour.'

'You are a slave driver. I was still deep asleep. Such a concert is very exhausting, you know!'

'I trust that, nevertheless come quickly. After a cup of coffee and a good breakfast you will feel better. We will soon find Peasley.' Astrid stopped the conversation and arranged her luggage. There was still time to ring Jenke before they would have breakfast.

Rio the Janeiro

With the shuttle, they arrived at half-past one in Rio. The airport lay still flat against the city and the bay and landing gave an anxious feeling. Astrid started immediately to find the security chief of the airport. Their luggage they had given in depot.

The chief of the security service was a small pedantic man. Nevertheless, when seated, this difference was not more obvious. Major Samaranta, thus he had presented himself, opened the conversation. 'What can I do for my charming guests? Have you lost something or have you been robbed?

'We have not lost something and neither been robbed! But the UN is robbed. Astrid nodded as if this increased the clarity for Major Samaranta.

'Ah... ha.., a joke, ha..., ha..., or must I take it seriously?'

'Of course, but we have not yet presented ourselves,' said Astrid. She pulled from her bag the UN document which had been given to her and gave it to Major Samaranta.

He bended over his desk, took the document and started to study it thoroughly. During reading the colour on his cheeks changed. After some coughing at the last page he stood up.

'My valuable colleagues, so I can call you. I'm naturally prepared to be at your service with all my capacities.' He hesitated realising the logic of his statement.

'But why are you coming here? I realise that a precious apparatus has disappeared, but I don't know why you are in Rio.'

'That can be rapidly explained. The theft has taken place around Christmas and we think that the thief has passed this airport. Can we have a look on the videos of all arrived passengers? We hope to recognise the thief.'

'Of course, with all pleasure. You can examine the videos in that room over there. Someone of the video service can assist you. Would you like to have some drinks; it might be a long afternoon for you because in that period approximately two hundred planes are landing each day.'

He stood up, but realised then suddenly that the ladies were at the wrong address. That thief might have arrived of course at the international airport. 'From where did your thief come from, ladies?'

'From Washington, Major.'

'Then it will last longer before I have your video disks and passenger lists, which must come, as it happens, from my colleague of the international airport. But since you are here, I will ask them to send these by heli. Do you have already a hotel in Rio? Yes? Then you can install yourself there meanwhile. Within two hours I will have the material.'

Astrid nodded. She started to like the little Major. How easily he could have sent them away. This airport was near the hotel they had reserved and the international airport was far outside the town.

'Major Samaranta, we will gladly accept your offer. It's extraordinarily nice of you. In two hours we will return.'

The Major beamed and caught the telephone to ring his colleague at the international airport.

So far everything ran as planned and Svetlana and Astrid took there time to regard the video pictures. After some hours they had examined all the videos of the flights from the US, but they hadn't discovered their thick friend, or rather there had been a lot of thick friends. For certainty they examined also the videos of flights which connected on flights from Washington or New York, but also that produced nothing.

'That Peasley was playing tricks!' said Svetlana. 'He might be in Australia!'

'Who knows..., but we won't give up. Let us first have a drink and then start all over again, only slower.'

The technician programmed again the same videos, but he projected these on a larger screen. Svetlana and Astrid peered to the arriving passengers who showed their passports at the customs. From each passport a picture was made that the technician projected in the corner of the screen.

'Why Americans look always so American. Whether they are black, brown or white, you can pick them up immediately.'

'Don't divert me, Astrid, look here to that man. He keeps his head down as if he knows that he is taken. Mr technician, can you repeat this part very slowly?'

The technician did what Svetlana asked and the thick man reappeared again. He had a hand bag in which there was to see a square object. His head was bald and Astrid compared it with the photograph which she had of Peasley. There was one who had been taken from above. 'Look at this photograph and at the picture on the screen. Bald is bald, but aren't these two pictures terribly similar? Can you show us the passport of this person?'

The technician let the man pass through the customs and the picture of the passport appeared in the right corner.

'It's him, Svetlana, look..! The photograph is the same such as that one we got from the US Security Service. You can read his name?'

Svetlana ran closer to the screen and tried to decipher the name.

'The name is not Peasley. It looks like Tvoorer or Trourer, I can't read it well.' The custom official moved then the passport under prerecording sensor. 'As a first name I see Benjamin. But the first two characters of his surname seems be a t and an r.'

'This is more luck than I had expected. He went simply to Rio. As an old veteran he should have known that we would come behind him. Or is that a trick, too?'

'Let us look at the passenger list,' spoke Svetlana. 'Look here Astrid, here you have him. He's called Benjamin Troover. We have him! Is that good? Let us go to our friend the Major, perhaps he can help us where to search further on.'

The two ladies thanked the technician warmly and went to the office of Major Samaranta. He showed himself very happy with the result as if he was entirely involved. 'Can I help you still further, ladies? With this name I can do something.'

'Perhaps, Major. How can we obtain access to the hotel registers of Rio?'

'Oh, very easy, ladies. Leave that to me. You take a rest in your hotel and I ensure you that we have tonight the name of the hotel where this Troover is lodged. I have friends at the police and that will help. You agree?'

Astrid and Svetlana couldn't do otherwise than to accept. This major was not only kind but he wanted to confirm his importance with this case.

'Would you be so kind to keep this as secret as possible. Peasley or rather Troover could have here some friends, and he shouldn't know that we are behind him. And believe us; he is notorious in his profession.'

'Don't worry ladies. I am as close as wax. Even the police force will not know why I want to check the hotel registers. I invent something about lost luggage. We will see us at approximately eight o'clock for diner in your hotel?'

'Excellent Major, thank you very much for your excellent help.' Astrid gave him a hand and followed Svetlana to the exit. By taxi they were rapidly in their hotel still in time to see the sun sink over the city.

Moscow

'See what's written here! "Send immediately a slipper of Peasley to us". What's that for nonsense?' exclaimed Professor Bolotnikov. He was on Saturday morning in his office to settle a number of subjects. 'What must Astrid and Svetlana do with Peasley's slipper in Rio de Janeiro.' He examined the e-mail once more with his secretary, but there was no further explanation. Bolotnikov caught the telephone and dialled the house number of Yuri Kaspadov. This one took almost directly the phone and saw his highest boss on the screen.

'What can I do for you, Professor?'

'Do you know something about Peasley's slippers?'

'Yes, indeed Sir. Svetlana rang me yesterday night at my mother and told me the complete story. They have found out that Peasley landed in Rio as Benjamin Troover and lived in a hotel in Niterói on the other side of the bay until yesterday. He has left and nobody knows his destination. The ladies are helped by a certain Major Samaranta, head security service of the national airport in Rio and he has proposed a search with a sleuth dog to trace where Peasley walked around in Niterói. Since his hotel room had been cleaned and reoccupied, the ladies want a strong source of smell of Peasley. Because they suspect that the house keeper might have cleaned all dresses of Peasley, they came up with the idea of the slipper. Could you contact Washington and have the slipper send by messenger to Rio?'

'You ask me something my boy. In Washington they will wonder whether the Russian President has nothing else to do.

But I realise the importance of the matter. Did your satellites have already traced the zwikker in South-America?'

'Not yet, but we have a chance of one on three if Peasley is using it. Over a week that chance is one to one.'

'Watch him, Yuri. I will inform Nicolai Yesin, perhaps he can be helpful with a magmobile and a driver. If you hear something, let me know it at once.'

'I'll do that Professor.' He felt himself lonely, in spite of the fact that he had spoken for half an hour with Svetlana yesterday. Svetlana proved to be enchanted with their progress, but everything would depend now on the slipper. As long as the zwikker was not detected, this was their only possibility.

Professor Bolotnikov called immediately Smith in Washington. It was there late in the evening. Smith was not yet to bed and the two statesmen discussed first a number of problems, whereupon Bolotnikov mentioned the slipper. It took some time to persuade Smith. 'I've had many crazy requests in my life, but this one is the best. They will get tomorrow morning that slipper. Furthermore still something?'

'The zwikker and terrorism, but that's a special topic for later, or rather for soon?'

'I'll think about that,' answered Smith.

Rio the Janeiro & Niterói

Astrid and Svetlana were the next morning disturbed at their breakfast by an American who asked whether they could legitimise themselves. Astrid got the inclination to become angry. 'What had this American to do with them.'

But the man insisted and mentioned the word "slipper".

'Oh, you are our good news bringer,' exclaimed Svetlana. 'I am Svetlana and this is Astrid,' whereas she took her passport from her bag. Astrid did the same.

The man studied the documents attentively, nodded and gave them a bundle from his briefcase. 'The slipper, ladies. In plastic packing. I will leave you now, you must forget that I've been here.' He bowed slightly and disappeared.

'At last a real spy, Svetlana. This man acts at least mysteriously. Did you saw him looking around as if he was followed? I hope he will return alive.'

'Do you think he is at risk?'

'Of course not, I can't imagine that Peasley has set up again a secret service. He must do everything on his own. And here he is no longer, that's clear. We will ring our Major after breakfast and then he can come immediately with his dog.'

Major Samaranta expressed his joy concerning the news that the slipper had arrived. He would pick them up at ten o'clock to go to Niterói.

In the car was not only the Major, but also a dark man who presented himself as Alduaje, Hermann Alduaje. His dog was a shepherd, called Bastje. The dog wagged his tail to Astrid and Svetlana and Alduaje let him smell their hands. 'Good people.'

'He does not bite?' asked Svetlana.

'Oh no, no..., this dog can be completely trusted as long as you are kind to him, isn't it Bastje?'

Niterói

Arrived in Niterći, they brought the dog to the hotel room which had inhabited Peasley and let the dog smell the slipper. Bastje withdrew first his head and looked at his boss with a 'must I?' Then he sniffed once more and left the room. He went to the elevator and below he walked into the hall. Then he went outside and walked in the direction of the inner town. After a couple blocks he stopped at the post office. After some hesitation he entered started to bark at the counter where packages were handled.

Major Samaranta asked the lady behind the counter whether she could give him information about a package which was sent by a Mr Peasley or Mr Troover.

'I'm not allowed to give you this information Sir, unless you have an authorisation of the director and he is absent.'

'Mrs, we're in a great hurry, would this help? The Major shifted a bank note in the drawer of the counter and showed his card of his security service.

'Only whether such a package has existed?' said the young counter lady hesitatingly. 'It remains between us. I have never done this, O.K.?'

'Splendid..., no word from us.'

'Who did you say? Troover or Peasley.'

'Indeed..., you've an excellent memory for names.'

The counter young lady typed the two names on her computer. 'Peasley,... no and Troover neither. Only if this fellow has sent something under cash on delivery, we make notes of the consignment, but that is not the case.'

'Do you know this man perhaps? He showed the photograph.'

'There are coming here so many men, Mr, I would not know, really not.'

Disappointed they left the post office.

'Is Bastje able find still more?' asked Astrid, 'for example where Peasley rented or bought a magmobile.

'Let us return to the hotel and walk in always larger circles with Bastje and see when he indicates a garage,' proposed the Major.

The search with Bastje had only success late in the afternoon. They wanted almost to give it up. They were tired and the January heat was heavy. On at least two kilometres from the hotel, Bastje barked. He pulled them to a garage where magmobiles could be rented or bought. Bastje had taken up the trace of Peasley at a place where he had stepped probably from a taxi, to visit the garage.

'Do you know this person?' said Major Samaranta to the mechanic.

'No Sir,' said the man, 'I don't know him. Why, what do you want from him?'

'This person, and perhaps he looked somewhat different, must recently have rented or bought a magmobile. Is that possible? A thick bold man with plenty of muscles.'

'Hey, Marguerita, come here. He waved to his secretary. 'Do you know you a fat man who bought last week a magmobile? Here's his photograph.'

'A thick man, with many muscles,' giggled Marguerita. 'I like strong men, but this man had a beard and the man on the photograph has been shaved.'

'That's it naturally' exclaimed Astrid. 'It's as we thought, Peasley did let his beard grow.'

'The name was not Peasley,' said Marguerita, 'he was called Troover, Benjamin Troover. But he did not look like a benjamin, ha... ha...!'

'Do you know where he has gone?' asked the Major.

'Mr Troover has paid the car cash and asked for a road map of South-Brazil and Uruguay. That's all I know. The man was not very communicative, muttered on y, but paid the price for the car which we asked. We could have asked 10% more. He was obviously in a hurry.'

'Thank you very much. Thus we can perhaps find our friend Troover.'

'Oh was he your friend? But I don't know more than I've told you.'

The Major, Astrid and Svetlana left the garage. 'You must also drive to the south, ladies, but there I cannot be of further help. If I you were, I would ask the help of a heli or of a fast transport vehicle. Perhaps you can trace this way his magmobile. Here's a folder of Peasley's magmobile. I got it from the mechanic as also his licence number.'

Returned at their hotel, Svetlana and Astrid thanked the Major very cordially and shook hands with Alduaje and Bastje. Svetlana got a lick from the dog.

'That we've done quite well,' chuckled Astrid, 'and what a nice man is that Major. He should have la decoration from the UN, don't you think so?'

'But what should we do now, have you some idea?'

'We will leave, but let us first have a bath and some rest. With time comes counsel. We must nevertheless inform home with a proposal. I first will take a cold bath. The heat is unbearable.'

'And I'll ring Moscow.'

'Again? But you are still young. My Jenke must wait before he sees me on the screen. Yuri passes faithfully the bulletins to Jenke?'

'I will ask him.'

Svetlana went to her own room. There she found a message that someone had a car with driver ready for them.

'They made already a decision for us. Yuri thinks of me, how dear. A weight fell of her heart and she tapped the secret number by which she could reach Yuri through Professor Bolotnikov. Yuri was already in bed, but he took the phone at the third ring.

'Were you asleep?'

'Slept...., yes I believe so. Is it you?'

Svetlana told him of their success and asked for suggestions.

'You will get a magmobile with an experienced driver, one from Interpol. Thus you can follow the track of Peasley. From the car you can remain in contact with us. As soon as we pick up Peasley's zwikker with the satellites, we will inform you. Have you your detector still?'

'Yes sweetheart, I have still everything. And I love you. More instructions?'

'Not directly, I love you too.' Yuri made kiss and broke the connection. Svetlana looked at his disappearing head and sighed. She had gladly spoken somewhat longer. She stood up and went to the bathroom. She let it fill up and stepped in it and started to relax.

She awoke with a bump by the buzzing telephone. Had she fallen asleep? She looked at on her watch. It was half an hour later. She left the bathroom and ran to her room.

The man on the telephone display looked sympathetic and reliable. 'Our driver, thought Svetlana.

'Are you Mrs Astrid Holthus? And have you found my message?'

'Mrs Holthus is my hairdresser and company, Sir. But I have received your message. I assume that you are downstairs in the hall?'

'Indeed. Can you and Mrs Holthus be in half an hour downstairs in the lobby? I see you require some time to dress.'

Svetlana startled, she had not realised that he could see her in her hastily arranged bath towe. 'We will be there. What's your name?

'Steve Cantonny..., I'm in the lobby.' He broke the connection.

Svetlana rang directly Astrid.

'You see Astrid, indeed with time came counsel. And that counsel is downstairs.'

In the hall they saw a broad shouldered man who approached them pleasantly. He knew them certainly already by face. 'You're Mrs Holthus and you're Miss Shornakova, Astrid and Svetlana, if I'm informed well.'

'And you are Steve,' said Astrid. 'Do you speak Portuguese?'

'It's my mother tongue, Mrs, or can say I Astrid?'

'Of course, and also Svetlana, isn't it?' Svetlana nodded.

'My mother is a Brazilian and I'm born in the neighbourhood. Afterwards we lived ten years in the US and returned lately. Here's my Interpol card and my CV. Of you both I've a complete file. How does it go with your future spouse, Svetlana? When do we leave?'

Svetlana blushed slightly, but gave no answer. 'It's none of his business,' she thought, 'playing prince charming. I've to watch out for him.'

'We can be ready in an hour,' said Astrid. 'Where we are going?'

'I don't know, you give the orders and I assist you when troubles arise.'

'Well then, we will drive southward and at each toll station you should check whether a magmobile with this number has passed. With a good tip they can possibly inform you.'

'Okido, ladies, over an hour I'll be here with the car and we can leave. You want to take still something special?'

'Yes, sandwiches and drinks. We can't stop always. Is your car large enough that we can sleep when you drive?'

'Sure. It's the newest model, with everything in it to radio and Internet. We can telephone worldwide and you can look at retransmissions to the threetel.'

'Oof, he likes chatting,' thought Svetlana. 'To impress probably. I'll warn Astrid, she knows perhaps how to handle him.' Aloud she said: 'You come along Astrid, so we can catch our luggage?'

In the elevator she pronounced her suspiciousness with respect to Steve.

'I've noticed that too, girl. Don't worry, I'll tame him. Men must always play "cocks" first, that's in them. Let them be "cocks" and if he continues to twaddle, you must simulate sleeping with a light snore. That's very effective.' She laughed and tapped Svetlana on the back.

'I'm curious what will happen next. So far it has gone better than I had dared to hope. That Major was as a matter of fact very nice with us. We'll send him a postcard.'

Thus chatting they came at their rooms, caught their belongings and looked whether they hadn't left anything behind. Svetlana tried just her zwikker detector, but there was no response. Peasley was probably far away and had the thing not working.

Steve held politely the doors open of his magmobile. He was obviously proud of the car. When driving, he started to tell about its radius of action and top speed. But Svetlana and Astrid looked only outside, the first softly humming. Coming on the highway to the south, Steve programmed the car until the first toll station and left the wheel to the automat. 'Can I be at your service with a cup coffee?'

'Deliciously, my boy. Such a practical thing I appreciate in a man. Please with a bit of milk. And you Svetlana?'

'Yes, me too, with some sugar.'

The coffee break diminished the tension and it proved that Steve knew a lot more then just driving. He told them about the strategy to follow for searching a needle in the hay stack without being noticed. Interpol was of course also busy and perhaps they would take underway contact. He expected a message that the zwikker had been detected by a satellite. It appeared that Steve was completely briefed. He was moreover cultural and played the German flute. 'Not too bad according to my critics. Look I've taken it along. I must practise each day to maintain my skill. Perhaps we can play together underway. I have music for a duet with me. You, Astrid, do you play something? I could find nothing in your CV.'

'I'm the best admirer of people who make music. You would do me a large pleasure to play together for me.'

In the meantime she thought about the dangers they could run. To find Peasley was still something else then to take away Peasley's zwikker. They had said nothing about that in Moscow. She had to trace the apparatus, that's all. Would there be a shade team ready to overpower Peasley? But if that team did not show up? And if Peasley would defend himself with the zwikker against them or "would retrain them"?

'How long we must travel until the first stop?' asked Astrid, throwing herself on practical things.

'Approximately four hours,' answered Steve. 'I propose we will sleep after our cup of coffee, which is almost ready. I will keep the guard.'

He didn't need to say that twice to the ladies and ten minutes later it was quiet in the large magmobile, except of the rustling wind.

16

What a joy to be able to free others from delusions.

Near Santa Dolores, south Brazil

Peasley had already arrived on his chosen place of destination. The climate pleased him; it was cooler and not so humid. In the rural hotel of a small village, he got all necessary information concerning sects which stayed in the neighbourhood. There was one who slaughtered in old testament manner sheep and lambs. Since each sect member was expected to be present at their worship, Peasley had selected this sect. Moreover he had something against maltreating animals, a weakness which one would not have thought of him. Although not being a vegetarian, he disliked killing and torturing animals.

He had loaded the batteries of the zwikker during the night and had packed these with the zwikker in a box so that only he could see the display. He had already analysed his own feelings for animals in order to indoctrinate those in the sect members.

He was happy that his magmobile was also built for heavy areas, since it appeared that the worship took place at an open spot in the landscape. Peasley could park his car on a small hill, so that he had a good overview. Some sect members walked around in long dresses. The service would probably start shortly. One seemed to wait for a signal.

The hullabaloo of voices coincided with the consonances of a trumpet orchestra at the centre of the open spot. Sheep or lambs were not in view. The mob was somewhat restless, until from the right side a horse and cart appeared on which a remarkable figure stood. Everything of ornament embellished him and Peasley thought he was a living Christmas tree.

The cart was conducted to the middle and the orchestra started to play louder. The sect members busted out in singing. They roared more than they sang. On a sign of the "Christmas tree" it became quiet and from the left side of the field walked a man in a white dress towards the cart with a lamb on his arm.

Peasley started the zwikker, aimed the sensor at the man with the lamb and took the spectrum and let at the same time the computer calculate the compensation spectrum. The part, of which he suspected that had to do with the worship and sacrificing animals, was clearly visible and he placed it in brackets. Afterwards he aimed rapidly the sensor at the man and pressed on the button "compensation" and the selected spectrum disappeared. The man with the lamb on his arm halted stand stiff as a rod and looked foolishly around. He did not seem to understand what he was doing there and for which reason he was dressed in a kind of robe. Peasley transmitted next his own bit of spectrum to the man and again the man got a shock. One saw that his eyes became clear and that he took a decision. He placed the lamb down at the ground and caressed it carefully. The lamb bleated softly. Some distance away a sheep answered. It had recognised the sound of its young.

The "Christmas tree" regarding this, climbed from the cart and walked to the man. Peasley could not hear what was said, but surely they were in disagreement. The "Christmas tree" wanted to catch the lamb, but the man took it again on his arm and ran away from the open spot. The "Christmas tree" climbed on a stone, levied high his staff and pronounced what was obviously a curse to the renegade.

Peasley found it now time to tackle the matter on a larger scale. He turned the sensor on wide angle. Would he be able to compensate all sect members in one time? He started again the zwikker. He had stopped it after the first test in order to avoid discovery by satellites.

The common spectrum of the crowd was somewhat confused, but the part which he had observed at the man with the lamb, was clearly visible. Peasley pushed the "compensation" button and aimed the sensor at the mob. Then he transmitted as fast as possible his own bit of spectrum.

Since he had not looked outside, he had not noticed the first responses. What he saw now was that the mob stripped off their dresses and walked away in all directions. Only the "Christmas tree" stood still quiet in the middle with his staff in the air.

'Just as a joke. Let me see whether it works.' Peasley put the sensor on sharp and gravitation and aimed at the staff. When the "Christmas tree" released his staff this continued to hang in the air. The man jumped frightened aside and looked astonished how his staff descended slowly to the ground. Peasley chuckled and turned out slowly the compensation.

'Ha.., ha.., they deserved that!'

Just now he realised how many sheep and lambs had been driven together on the other side of the field. Would all these have been sacrificed? Peasley shivered over his heavy body at the idea. The sheep were released by one of the men without any further consideration. Peasley smiled and tapped admiringly on the little box that had caused all this. He had no regret what so ever.

'Authorities would never have done such a test,' he muttered, 'they would disapprove it under the cloak of religious freedom. And what about the animal rights?'

He stored the zwikker away, looked at the sky as if he could detect any satellites and drove back to the hotel. There was a great upheaval at the bar. Ex-sect members seemed to be very thirsty. The landlord glowed of pleasure because of the extra customers.

'In former days those fellows would never came here,' he said, when Peasley asked what was the matter. 'They have hung their dresses to the trees. All sheep and lambs have been released. You understand that, stranger?'

'Perhaps their leader has seen the light,' answered Peasley with a smile that slowly developed into a broad and thundering laugh. The bar visitors looked first frightened, but Peasley's laugh worked so contagiously that most of them joint laughing. 'How, how, ha, ha..., is that laughing,' he hiccupped, 'I haven't laughed so much in years. I didn't know laughing still existed. I give a round to you all. Give them something real Brazilian, no import.'

Peasley got an applause and a number of men came sitting at his table.

'Mister, that was a crazy afternoon, believe me. I'm from that sect. I can't even more believe why. How do you find that?'

Peasley did not answer, smiled to the man and shrug his broad shoulders. He knew now enough, the test had been a success. Now the larger work. There was much to be improved in the world. Tomorrow he would try the zwikker by means of the threetel. There should be some political event in Montevideo, Uruguay, he had seen on the news. It was supposed to be broadcasted life tomorrow morning.

What he wasn't aware of was that one of the satellite detectors had captured the signal of his zwikker. Fortunately he didn't know, because for the first time since his departure from Washington he had a healthy sleep and awoke late the next morning.

The event in Montevideo was an electorate meeting. Sitting in front of his threetel, Peasley listened to the speaker announcing the meeting. The public reacted with moderate enthusiasm. Obviously they were awaiting the big chief. There were banners with slogans for more power to the people and away with the "intelectua". That was certainly the dominating class. Peasley had the zwikker ready for one fast detection and compersation. The best moment seemed to him during a speech and when the people applauded.

An orchestra played. Suddenly there was some randy at the entrance and a group of men and women entered, dressed in the colours of the party. In their middle was the leader, who gave hands, waved and embraced women.

'The usual theatre,' thought Peasley, 'what's only lacking is kissing babies. We will get them astonished soon. Or perhaps not, because they will be compensated. The festival will start now.'

The Chairman of the meeting announced loudly the leader, who strode up to the podium. Peasley had admiration for the style of the man. He had charisma.

This was also proven when he started to speak. He brought his speech clearly, spoke with short senses and ensured regularly a vocabulary climax whereupon he was loudly applauded. At the exclamation "away with the intelectua" the applause became an ovation. Peasley started the zwikker. He saw a spectrum with in the right part high peaks with a number valleys. With the computer he selected these peaks and pressed on the button to calculate the compensation pattern. On the screen an identical mirror picture appeared.

The leader continued in the meantime and the public was hanging at his lips. The commentator kept its mouth. Peasley aimed the sensor exactly at threetel-screen. When the public applauded again, he pressed on the compensation button. He saw on the screen the compensation spectrum slowly decreasing until a message appeared with "Emission completed, you want another one? Y/N" Peasley pressed on N-button and the screen went empty.

He paid attention to his threetel, where the applause had struck dumb. The leader looked foolishly around as if he was searching for his text. But there was no text since he had spoken by heart. He focused again on the people and said something like: 'I have said...' left the podium and the hall. The people followed him without interest and started to talk to each other.

'This is really something of world influence. I could have indoctrinated them with something else, but for the moment it is already a terrific result. That Kaspadov has really invented something.' A deep satisfaction rose in him, more than he ever had felt during its previous career.

'Oh my god, I must still turn off the zwikker. They might have traced me. I must leave immediately. I will go to Uruguay to see the impact on the people who watched the threetel.'

He wrapped up the zwikker, paid is hotel and mounted his magmobile. Within ten minutes he was on the motorway and programmed his car until the border.

He had been correct. They had indeed traced the zwikker. Two of the satellites had passed over and it did not last long before Kaspadov and colleagues had determined where the zwikker had functioned.

Yuri did not hesitate to inform Professor Bolotnikov and within half an hour President Smith and Dolores Guerrero were informed. It was decided that Interpol in association with the two ladies would go behind Peasley. Dolores arranged with the President of Brazil about the use of a Brazilian heli.

The telephone in the magmobile of Astrid and Svetlana buzzed when they had together a drink and were travelling through a splendid pampas region. Astrid took it. 'Ah, Victor, you have some news? We are driving south behind Peasley. He is some days in front of us, but the toll services do help us well.'

'Is Svetlana there too?'

'Yes certainly, Professor, here she is.'

'Well youngsters, listen attentively and take a map that goes to the Uruguayan border.'

'I give you the coordinates where we traced Peasley's zwikker. Repeat it for certainty.' Svetlana complied with the request and saw Bolotnikov nodding.

'Peasley has used twice the zwikker. Yesterday very shortly and approximately two hours ago, for ten minutes. You must drive as quickly as possible to that spot and examine what he has done there. We will send also a heli of Interpol. However I don't think Peasley is still there. If he has gone indeed, you must continue to trace him. You should under any circumstances, and I repeat under any circumstances, overpower Peasley. He's a dangerous man. You got that?'

'Yes, yes, we've everything. As soon as we know something we let you know.' Astrid waved and Bolotnikov's picture zoomed away.

'Why do you call him Victor and why is Svetlana addressing him with Professor?' asked Steve.

'Oh, that's a long tale which I will tell you later. Check now rapidly in how much time we can arrive at the indicated spot. It has no sense to stop at the toll stations, which only cost extra time. And you Svetlana, can make again some coffee?'

Astrid felt herself radiating. The tension did her good and at such moments she almost forgot Jenke and the children. 'If we only could succeed now. Wouldn't that be marvellous?' she thought.

'I think we can be there tonight,' spoke Steve. 'Look, we have a motorway until there from where we simply continue on our wheels. Then I can show you what kind of a good driver I am.'

'Then we have little time to think what Peasley could have done. Afterwards we will bump with our head against the roof, I've understood,' looking at Steve.

He laughed widely and put up his thumb. He wanted just to start a discussion on his driver's art, when Astrid said to him: 'Steve, please look at all large and small roads to and from the indicated spot, because we probably have to continue straight on after our arrival. This gives us time to investigate the local news on Internet.'

Astrid searched systematically all press bulletins of yesterday and today of that location until the border. Because she pressed the simultaneous translator, she got the original reports next to the English translation on the screen.

'The press is pretty local, you think too Svetlana? Nothing about international matter. No wonder Peasley selected this region. They've never heard of the zwikker. Hey, there seem to be quite a number of sects here. Give me the C.V. of Peasley whether he had interests in sects. Perhaps he wants to turn the complete world into one sect. How about that, everybody dancing in pink dresses.'

Svetlana read the C.V. thoroughly, but couldn't find anything concerning sects. 'He doesn't seem to be a sect adepter; on the contrary, he rather hated non-American religious doings.'

'Wait, yesterday a sect kept their annual meeting where in old testament way lambs and sheep would be sacrificed. How stood Peasley compared to animals?'

Svetlana passed again through the C.V. 'Nothing concerning animals, Astrid. He had no pets.'

Astrid wrinkled her face. Her intuition said her there had to be an indication. 'Look once more, is he vegetarian?'

To the C.V. was an appendix in which his daily habits had been described.

'No, he is not a vegetarian, but he consumes only flesh of adult animals. Therefore never lamb - or calf flesh. How about that?'

'Perhaps, no lamb flesh therefore. And that sect sacrifices lambs. Let us screen the press again about lambs. You never know.'

Astrid was for almost an hour investigating the bulletins. She thought: 'That press freedom for everybody might be beautiful, however, it takes much work to read all those reports. The words lamb, sheep and sect produced nothing.'

With a tenacity which Jenke knew well, she continued until she suddenly shouted. Steve shot straight from his chair and Svetlana called: 'What have you found?'

'Look here. There were complains that all animals from the sect members were released in the woods, while the sect members had disappeared. These animals were the sheep and lambs which they wanted to sacrifice. And they haven't done that.'

'Suppose I was Peasley,' continued Astrid, 'I would also use the zwikker to prevent such a sacrificial feast. If he has done that, he is growing in my esteem. He is not the iron man as he is known at the US Security Service. Svetlana we must search further, because he has used this morning again the zwikker and for a longer period. But for what? There is further nothing more, and the bulletins of this morning haven't yet been presented. Nevertheless an indispensable instrument, this Internet.'

'We will leave the high way in ten minutes,' intervened Steve. 'Fix everything well. There are belts enough.'

His remark brought Astrid and Svetlana back to reality and they stowed their luggage and equipment. After ten minutes Steve started the large engine which operated the wheels and drove away over the traditionally paved roads to the west. He had calculated that it was approximately ten kilometres to the village, of which Professor Bolotnikov had given the coordinates. There arrived there at the only rural hotel. They asked for three rooms, but told they might possibly leave again in the night. Astrid made a chat with the receptionist and showed the photograph of Peasley which they had updated with a beard and a moustache.

'That's that Mister who has left this morning! But he has only a short beard, no moustache. Nice man. He came the day before yesterday and has explored the hills. He's an amateur bird lover. We showed him the best location where he could photograph. But he told us he registered only bird singing. He had a recorder with him.'

'Have you listened to his recordings?'

'No..., he had, however, promised it, but this morning at eleven o'clock he suddenly came down and left within an hour. A good tip, nice man gave.'

Astrid got the hint and put some money on the counter. 'We're from the same University. We should meet together for a research. Do you know when he returns?'

'Sorry, I don't know. Personally I don't think soon, because my husband met him when he turned up the large motorway to Uruguay. I think he has gone to Montevideo, because I saw him at breakfast examining the map of Uruguay. Perhaps he has forgotten your appointment. I'm not surprised, because he also forgot to let me listen to his recordings.'

'What do you know of that sect, which were uphill yesterday?'

'Yes, what do you want to know..... eh?' spoke the receptionist.

Astrid placed some more money on the counter, which the receptionist snatched away very fast.

With her head bent forward she whispered: 'They became stone crazy. They came first in procession, dressed in long dresses and followed by a herd of sheep and lambs. Never a drink at the bar. And that leader! Apostle they call him, who is entirely strange. But yesterday, you can't believe it; those fellows came in without dress and ordered drinks. I had good day. They paid well.'

Astrid knew enough, she caught her case and went to her room.

She took a bathe and made up her face to feel herself again a bit normal. Even with rustling magmobiles travelling is tiring. She had soon taken a decision. Why waiting until tomorrow? Svetlana should first ring Moscow.

After an hour she came down. They had agreed with Steve to dine in the hotel. She found Svetlana in the restaurant, but Steve was not present.

'Where's Steve?'

'Certainly with his car,' replied Svetlana. 'I contacted Yuri, and he said there is no further news. We must decide by ourselves what to do. What do you think?'

'I hesitate between travelling tonight or tomorrow. Let us find Steve and I will tell what I've heard from the lady in the lobby.'

They found Steve in busy conversation with three unknown men. Further away stood a heli of a type which Astrid had never seen. A group of children ran curiously around.

'He Steve, come here?' called Svetlana.

Steve turned around and waved the others to come closer.

'These are two colleagues of my service and a Brazilian pilot. The two Interpol members gave Svetlana and Astrid a hand, the pilot bowed slightly.

'Nice to meet you,' said Astrid. 'We've to discuss something with you. Can I invite you for a drink in the hotel?'

The three men and Steve nodded and followed her. The pilot preferred to stay. 'There are too many curious children here. I prefer not to leave my chopper alone.'

Astrid stopped Svetlana a moment and whispered: 'Don't tell everything, because it does not seem good to me that these men catch the zwikker. Suppose they want to keep him for themselves, or they will joint Peasley. Let me do the talking.'

The men followed her and she ordered fruit cocktails for them. Herself she took a coffee. Svetlana took nothing.

'Gentlemen, I suppose you were sent to this spot by the Interpol. Is that correct?'

'Yes Mam,' answered the senior. 'These coordinates were given to us and the Brazilian security service gave us a heli with a pilot. We should receive further instructions at this place. Are you the one who will give these instructions? What is your function in all this?'

'If you have just a look at these papers, you will know that Svetlana Shornakova and I, Astrid Holthus, work for the Secretary-General of the UN on a secret task. Our task is to trace someone of US Security Service who has disappeared with important UN documents. His name is Benjamin Troover and he has a magmobile with this number. Yesterday he was in this hotel, but he has left this morning. He cannot be too far and my request to you is to find him for us so we can go by car behind him. Which direction he has gone I don't know. When you fly in a spiral you should be able to find him. As soon as you notice him, you pass this on to us and we will finish the job. It is a black carriage of a type which Steve can describe to you. I suggest you eat something and start flying before it becomes dark. This is our phone number. As soon as you get contact with us, you type the following code, so that we can speak freely. You take under no circumstances action to stop the carriage of Troover. Only trace him. Don't let him suspect that you follow him.

'I fear Mrs Holthus that I must have an affirmative of my boss for this task. Can you arrange that?'

'No sir, this is not possible. You have been told to receive here your instructions and those you've just received now. Nothing can go wrong for you when you only fly around and pass your findings to me.'

'Perhaps you're right, but if I surpass my orders, I expect you will cover me.'

'Of course. But enough chatting, please leave as soon as possible.'

The men took some sandwiches and disappeared to the heli which had got company. Besides about forty children, stood the local police force. Astrid saw them heavily gesticulating and shouting. Obviously the heli might not have landed there.

After a few minutes the very nervous Brazilian pilot entered and said that they should leave immediately unless the heli would be confiscated. 'In an hour we will send you a message if we have had a result,' shouted the oldest Interpol man from the starting heli.

Astrid waited until they were far out of sight. 'Steve prepare the carriage, we will not sit here waiting. I bed that Peasley is not yet very far south, so the closer we are when the heli traces him the better. If not, we just have bad luck. Come Svetlana we will catch our luggage.'

Above in their room Astrid said: 'I've something to say to you. From the receptionist I know that Peasley has worked here successfully on a sect. They were holding a sacrifice feast where they would slaughter sheep and lambs. I could make up from what she said that the sect members had changed entirely in behaviour. Later she saw that Peasley studied a road map of Uruguay and I suspect that he has gone there. If he has crossed the border those fellows in the heli cannot fly there. This gives us the chance to beat them. I want to catch the zwikker by myself, because I don't trust these Interpol fellows. Once knowing its power they might come under the spell of it. We must be fast, because Peasley has used the zwikker this morning again.'

'But now you go against the orders of Professor Bolotnikov!'

'I realise that, but he couldn't have foreseen this situation. Moreover it is possible that we can obtain the zwikker from Peasley in a gentle manner. He seems to have weak sides in his character, like respect for animals. Although, I wouldn't call it weak but rather strong. For this reason only, he is entitled to a merciful treatment. Those Interpol fellows would positively proceed to a hard treatment and who knows with a negative result. In such a situation Peasley might be the strongest. You understand now why I play it this way?'

'Yes, you're right, but Steve, must he be informed?'

'No, no, certainly not. I don't know which side he would choose, that of Interpol or ours. If he doesn't know our plan he will help us, since he thinks we simply gamble where Peasley might be. Come, let's go. I will settle up with the receptionist. She will think, what strange people. Oh, yes, Steve doesn't know either of your pocket detector. Perhaps we have to use it tonight or tomorrow. Don't show it to him.'

Steve, who obviously was used to strange decisions, reacted very simple: 'Where do we go? South, North, East or West?'

'I bed for south. And if we are fast, we can be tonight at the border of Uruguay. The motorway goes to there, isn't it?'

'No problem. As soon as we are on the motorway, we can have a nap and arrive even fresh at the border.'

It thus occurred and during some hours they rustled along the highway to Uruguay. At the last toll station for the border Steve found out that the magmobile of Peasley had indeed passed and was approximately three hours ahead. They passed for this reason the border and arrived late in the night near Montevideo. Underway they had a last contact with the heli and Steve had communicated that they were on the track. The heli could return to its basis.

Montevideo

'Oof,' they said when they arrived in a suburb of Montevideo. They had chosen the first available motel 'Oof, what a day. I like adventure, but this was a complete jolt.'

Steve had withdrawn in his own room and Svetlana and Astrid were together in a double room.

'Let us first test your pocket detector. Perhaps something may happen tonight or tomorrow morning. Ring also Yuri if he has news.'

Svetlana connected the pocket detector to the threetel of the room, but the screen remained blank. 'No activities within two kilometres from here.'

'Keep it on, Svetlana. Does it have a warning signal?'

'I believe so. Yuri mentioned something like that. I will look in the guide. Yes, press this button and a bell will ring when the detector measures something. I will now ring Yuri?'

After several attempts Yuri's face appeared. 'Everything O.K. with you? I see on my telephone display that you in Uruguay. How's that possible?'

'That's a long tale. We will tell that later. What we want to know whether you have measured some activity again. Ring immediately when that's the case. Is further all well with you? We are a bit tired from the long ride, therefore I stop now. See you soon, darling.'

Svetlana made kiss movement and broke of the connection.

That night, as far as they could speak still of a night, they slept profoundly. Only at ten o'clock they woke up. The screen of the threetel was still empty, it showed no activity.

Steve was already downstairs; he checked his magmobile and the morning newspaper which he had bought. Astrid approached curiously and looked over his shoulder to the bulletins in Spanish. Soon her board computer would show these bulletins in the English translation. One thing took her attention. There stood "Election meeting was a large disappointment."

'This might be caused by the second emission of the zwikker? But then Peasley was still in Brazil,' she thought.

She went to the magmobile and started the computer. The report was already available and she read: "Enthusiastic election meeting fell suddenly dead. In the middle of a glowing speech of the well-known politician, constantly interrupted by applause and cheers, a silence had fallen. The speaker just left the podium and walked away, a foolish public leaving behind. At inquiry nobody had a declaration on what had happened and they didn't even know why they had been so enthusiastic. It was not a case of massive amnesia. It seemed to be some negative mass psychosis. One stood for a mystery."

'This cannot be else than an action of Peasley, but how?' she thought. 'I should have better listened to the explanation which Jenke and his friend gave me in Washington. I will ask Svetlana, she knows perhaps more.'

She walked rapidly back to the hotel and found Svetlana in their room. Just when she wanted to put the question, a point blinked on the screen of the threetel and a bell was ringing. 'Astrid!' called Svetlana, 'look, Peasley has started the zwikker and he must be very close by. If we wait we can see how far he's away.'

At that moment the telephone rang and Yuri came in view. 'Svetlana, a zwikker has been identified. Look, I aim the camera just as at the map. It must be near you.'

'We have detected him too. Stay on the line, Yuri, I can say how far from here. The sensor is still counting. Look he has finished, it's only 300 meters. But that's very near!' called Svetlana enchanted.

'Be careful, Svetlana. I cannot help you with your decisions, but please be careful.' Yuri showed a concerned face. 'Must I ask for help?'

'No, not yet!' interrupted Astrid. 'Don't do anything and trust us. I have a plan that has a chance of success. And now stop, because we will be very busy.' She particularly didn't want intruders.

'In which direction is it?'

'Exactly in that direction if you look out of the window. It must be that motel there on the other side of the road. What shall we do?'

'We go there. And remember, I do the talking,' laughed Astrid. She felt the tension rise and this made her happy. This was adventure. That he should be a dangerous person, she didn't want to believe. Also a fat man with many muscles can be approached.

Svetlana and Astrid went down. Steve was not there. He needed permanently something for his car.

Arrived at the motel they saw the magmobile with the well-known number. Peasley was indeed in the neighbourhood. Svetlana had taken along the pocket detector and it gave a strong response. Since it wasn't connected to a threetel a flash light indicated whether a working zwikker was near. The strength of the light was reversely proportionally with the distance and as they came closer to the motel the light became stronger. Fortunately Peasley had not chosen a hotel because then they had to pass a reception. Now it was possible to go without any problem to his room. They went to the room at the angle of the motel where the magmobile was parked. The sensor flashed strongly.

Astrid pressed the bell with a pounding heart. They heard rumbling in the room and at the same time the flashlight of the pocket detector extinguished. Peasley had turned the zwikker off.

The door opened and in front of them stood the man they were after. The first impression of Svetlana was that the beard didn't suit him. It made him even heavier. Astrid took the word and spoke: 'Mr Troover, if I'm well informed?'

'And what matters you...?' was the answer.

'Not your name, but we are from the county health control department and we check all foreigners who come in Uruguay. We are afraid of the extension of a new epidemic sickness, called "Horseway", which is easily spread by physical contact. It is a virus sickness. Not deadly, but very annoying. The symptoms are....'

Peasley interrupted her: 'Say what I must do. I've little time for your explanations. Make it short!'

Astrid understood why this man had brought it once to a high function. She should not exaggerate otherwise he would smell danger. 'We must only inspect your room with this equipment; it is able to discover a contagion. My colleague has the apparatus in her hand. If you would be so kind to go just outside for some minutes, so we can execute our task. You are the last of the foreigners in this motel which we have to check and we must still do the hotel on the other side of the road.'

To her surprise Peasley stepped aside and kept the door open for her. On the table she noticed something that probably was the zwikker. 'What is it small,' she thought. 'Must certainly also be connected to the threetel.'

'Svetlana, can you hand over the sensor, please, so the measuring can be settled rapidly.

Astrid caught the sensor and simulated to search in every nook and corner. Under the bed she pressed the test button of the detector and the control flashlight started twinkling. She let out a cry.

'Mr Troover, I measure something. Whether this is a contagion which you have brought with you, or one already there I have to check. Can you please move outside and I will check you there.'

Peasley, looked slightly worried and suspicious, but attended to her request. Astrid followed him and closed the door of the motel room behind him. Then she went with the sensor, of which the test button was off, to the thick man and checked him over his body. 'No, you yourself are not contaminated, at least not yet. How long are you here already?'

'I've arrived yesterday evening?'

'Anyway Sir, we have to decontaminate your room and you have to take temporarily another one which is not contaminated. I suggest you go to the lobby and ask for another room, which we will inspect before you enter. You must leave your luggage in your current room. We will call our disinfection team and you can recover your things this afternoon. Please give me your key and when you go to the reception we will wait here in front of your door. I see that there are several rooms free.'

Peasley hesitated. He didn't show his distrust, but they could see he reflected somehow on something. Because he obviously didn't expected female agents on his trail, he muttered: 'O.K. ladies, wait here, I'm back in a minute.'

He disappeared around the corner. Astrid made a jump as a cat in action. She put the key in the lock, opened the door and caught the zwikker from the table. Outside again she ran for the corner of the motel and put down the zwikker under a bush. She had to take the risk that someone could steal it there. She returned to the room and stayed together with Svetlana until Peasley returned.

'I got the room next to mine, ladies. You want to inspect this one?'

Svetlana remained outside and Astrid did as if she checked the room thoroughly. 'This room is infection free, Sir. You can take it safely. You have perhaps enough reserve in your magmobile to endure a couple of hours? We hold back the key and you can expect us within an hour.'

Peasley muttered in himself. They could see that he didn't like it, but he couldn't discover any irregularity in what Astrid had asked. He looked still through the window of his old room, but since he could not see the table, he gave further no response.

He entered his new room and Astrid and Svetlana went around the corner to take a breath.

'Oof...,' said Svetlana, 'you will give me once a heart attack. I almost suffocated. Fortunately I have a good training as violinist to overcome stress. But oof, especially when Peasley looked through the window.'

'Come, we have the zwikker. Quick to our hotel and then an axe. I don't trust that thing as long as it can work. Afterwards we take the remainders along as evidence. No... not that side. Suppose he looks out of the window.'

They sneaked with a large arc around the motel to another street and walked back behind a hedge to their hotel. Arrived in their room Astrid put the apparatus on the ground and battered it with a chair. The zwikker was not resistant against this violence. After three slaps it succumbed and fell in pieces. The component of the gyroscope gave a large thud. Svetlana was clearly scared. But practical as Astrid was, she caught a case and slid all components in it, or rather what was left.

'So, we made it. Peasley is certainly out of training. If it is true what they said of him, he should never been cheated this way.'

'But must he not been taken in custody?'

'We should not interfere with that, you remember. So we won't do that. But I want nevertheless inform him of the destruction of his zwikker. I will show this part to him, and then he knows enough. Perhaps it is for him a turning point in his life. I believe he has also his good sides. You will come along, or you prefer to stay here?'

'To be honest, I rather stay here. One more thing and I will start screaming from tension. I haven't such a strong constitution as you.'

'Just wait until you must calm two whining children. Warn Steve if I am not back within half an hour.'

Astrid looked in the mirror if she was sufficient good-looking and walked along the same turning to the motel. When she turned the corner of the motel she saw Peasley standing on his toes in front of the window of his old room. She tapped him on his back with as a result he jumped high up and came down with a large boom.

'Mr Troover, can we discuss in your new room the results?'

'You scared me..., is that your manner of working? Strange I would say.'

'It was too a temptation Mr Troover, I've done that frequently with my husband. It's improving his soul, you know.

And perhaps your's as well. Come let us go inside. I can sit there?'

Peasley shuffled along a chair and bumped into a fauteuil. 'What do you want to tell?'

'May I call you Tim?' asked Astrid.

The reaction was one of complete stupefaction. Peasley's mouth fell open and he looked at her unbelievingly. Blood pulled away from his face. Then the colour returned gradually and he recovered something of his former pugnacity.

'You know me...?'

'Yes Tim and you can call me Astrid. I am the woman of a colleague of Ilja Travenkov, you know, the man who has stolen the plans of the zwikker in Moscow for you. My colleague here is the fiancé of Yuri Kaspadov a name you know as well.'

'And what do you want from me?'

'Nothing. Nothing. The zwikker, with which you saved the life of sheep and lambs in that village in Brazil, I've smashed to smithereens. Look, this you can take as a souvenir.'

Peasley had leaped as being stabbed with a needle. Astrid had not thought that a thick man could react so rapidly. He jumped up and wanted to catch her, but Astrid looked him deep in his eyes. She still thought of her dream that she should have learned karate. 'Tim, don't act crazy. I'm here to help you. There is an army of Interpol ready to lock you up for ever. You must know what it means to trace someone important. They always get him. I say always!'

Peasley blinked to overcome his anger. 'And that you are telling me just simply. Do you know what you've done?'

'Yes, I know that very much. I think the zwikker is a danger for the operator because he wants to predominate other people. Your two tests of the previous days were still reasonable innocent, but you were busy again with the zwikker. What were you measuring? That interests me.'

'You are a strange woman, Astrid,' said Peasley, who had regained his self-control. 'I was measuring my own spectrum. I wanted to know whether it was still the same as in Niterói. But how you've found me and where is Interpol?'

'Svetlana and I have found you sooner than Interpol, although our driver is from Interpol. If you had remained in Brazil they would have got you already. One can find you within some seconds if you start the zwikker. The system of satellite detectors is optimum and Svetlana has a pocket detector. Moreover you've seen that detector already, we used it to trace that virus. Sorry, but I wanted to prevent that they have to kill you. That would probably happen when Interpol would have found you. Further I don't want that the zwikker falls intact in their hands, because then the complete story will start again. I think that a man cannot resist the apparatus. A woman perhaps also not, but they have less inclination to predominate others and therefore are longer resistant against seduction.'

Peasley had listened to Astrid without interrupting. He had never encountered such a woman. At the US Security Service, however, there were female agents, but he had seldom contact with them.

'And now, Tim, I give you one hour time to disappear. With that I probably oppose the intention of Interpol and the UN, lets stand President Smith, but Svetlana and I were not allowed to arrest you under any circumstances. You are a dangerous man, they told us. Thus we won't do that. And what that dangerous concerns, I've seen that you have also a good heart. Let that speak more often and you will go well. You have one hour, then I must inform my driver and then the complete Interpol will know it. You have probable enough talent to escape. Here's the key of your old room. Perhaps we see each other once again. Good bye. And search for a nice woman.'

Peasley was too staggered to bring out a word. Of perplexity he forgot to answer.

'Oh, Astrid, I was so worried. I was exactly planning to come after you. You stayed away as much as twenty minutes! And I cannot find Steve, who had left with his magmobile.' Svetlana stood in front of the hotel and greeted Astrid who approached her in a kind of trance.

'What happened? You look so absent. Has Peasley done something to you?'

Astrid slowly shook her head. 'No, but almost. He stood there to seize me at my throat and I threw all my strength in my eyes to hold him back. Only now I realise what kind of a risk I've run. I think that next time I be "better a frightened woman than a dead woman". I still tremble.'

'Come rapid to our room and tell me everything. There is no more danger? He will not come here?' asked Svetlana, looking frightened over her shoulder to the opposite side.

'I don't think so he was taken absolutely by surprise. I've given him an hour to disappear and have said we will afterwards inform Interpol. In spite of the fact he had me almost at my throat, I have some compassion for him. It is such a lonely man.'

'Say, you'll deferd him already? I'll be glad when we will be safe again in Europe. To play hero is nothing for violinists.'

Astrid closed the curtains and glanced through an opening to the opposite side. Peasley was still in his old room. She saw some slight movement. The magmobile was still there, too.

Only after five minutes they saw the door opened and Peasley appeared with his luggage. He threw the trunks in his car, started it and drove to the head building. There he stopped for a moment, probably to pay the bill, and drove away.

'He has shaven his beard, you saw that?' called Svetlana. 'That's why it lasted so long before he left.'

'Ring now Professor Bolotnikov, Svetlana. We must inform him and ask for instructions.'

It lasted this time at least fifteen minutes before she had him on the line. 'What's the matter Svetlana?' asked Professor Bolotnikov when he saw who he had on the line.

'I've important news Sir. Look, here I have the rests of the zwikker of Peasley. Entirely in small parts. And we're still alive.'

'Can I hear the complete tale in correct order? And, devilish, where is Peasley?'

Astrid took over and told the story from their departure from Brazil. She finished with '... and then I gave him an hour to disappear. As a matter of fact there was nobody in the neighbourhood to arrest him. Steve was gone and what can two weak women do under such conditions?'

'Don't look so innocent, you call yourselves weak women!' exclaimed Victor. 'I will warn rapidly Interpol and either they go after Peasley themselves or give the job to the police force of Uruguay.'

'Not yet,' interrupted Astrid. 'I've given him one hour and that expires only in 20 minutes. My promise is also yours.'

Professor Bolotnikov frowned. Astrid saw that he hesitated and reflected on the impact of her promise. 'Well, Astrid, I will respect your promise. Finally we owe you that because you have made the zwikker harmless. But he gets no more time than you promised. And what concerns your stay there, go to the airport and take the first flight to Moscow or the one with the quickest connection. I expect you tomorrow in my office. But before I leave you, my congratulations for your result. I will be glad to see you again in good health.'

Whereas Victor disappeared in the usual way from the screen, Steve stormed in their room without knocking at the door. 'Quick! Do you know you who I encountered? Peasley in his magmobile. He seemed to be in a hurry. He has gone to the city and if I alarm Interpol, they might be able to catch him. Quick, take your luggage and off we go!'

He wanted to catch the telephone, but Astrid put her hand on his arm. 'Don't do it, Steve. Everything is already arranged. You've nothing more to do. We just spoke with Moscow and they put everything in functioning. And look here on the floor, there are the remainders of Peasley's zwikker.'

Steve stood gaping as if he saw men from Mars or rather women from Mars. 'What ho.. Say you ….what ho? This was a zwikker? And Peasley gave it to you?'

'Something like that... yes. Svetlana will tell you the complete story. I've just explained everything to Professor Bolotnikov in Moscow. And when Svetlana is ready, you can bring us to the airport? We should return to Europe.'

Svetlana told it as good as possible. At the part concerning the promise of one hour delay, she saw Astrid laying her finger on her mouth. She omitted it therefore and continued until the conversation with Professor Bolotnikov.

'I can't believe it. And if Peasley had attacked you I wasn't even in the neighbourhood. Don't tell this to anybody, would you? It may cost me my job. Finally I had to assist you physically. Why haven't you waited until I returned? He could have assassinated you!'

Astrid did not react on his outburst. He was rather hurt in his self-respect then being happy with the result. But such kind of responses she had experienced earlier with men.

'We are ready in five minutes, Steve. Don't take contact with your bosses than over an hour. In that time they have had instructions from above. We must grant Professor Bolotnikov some time. Under no circumstances try to catch Peasley. Getting the collaboration from the Uruguayan police takes more time than when this is regulated from above. You must first bring us to the airport. She did not contradict the promise of one hour delay for Peasley. Bringing them to the airport would be more than enough.

Steve muttered, but made no further opposition. Within half an hour they were on their way to the airport of Montevideo and an hour later they sat in a plane to São Paulo where they had a connection to Moscow. The remainders of the zwikker were neatly packed by Astrid in a small box which she had in her hand luggage. It was finally their evidence material and should not fall in strange hands.'

Moscow

Arrived in Moscow they found Jenke and Yuri waiting at the terminal. Svetlana jumped Yuri at his neck and Jenke closed his wife firmly in his arms. 'What have you both done?' Jenke said to Astrid. 'Oof...! How I'm glad you are back safely. And we go home already tonight. The children will welcome you with a triumphal arc. They think you saved humanity. I had them just on the line and I've told them everything. We must go now to the President and then you are free.'

'Oh, Jenke, what is it nice to see again your blond hair. I had a terribly tense time. Deliciously it was, but at the end I've been quite scared, I must admit. Some rest will do me good. You go along to the President?'

'Yes, yes. If the President tries to reprehend you, he has to face me. Yuri told me that Peasley has gone up in smoke. They have found its magmobile in Montevideo, but he himself left no trace at all. The complete Interpol is behind him.'

'That's good! I like it he has escaped. Finally he has been reasonably correct with me. Fortunately he was not yet much under the influence of the zwikker. I hope that he stays untraceable.'

'Astrid, you remain a surprising woman. I wonder sometimes whenever I will annoy myself with you. I suppose never. But now first to the President. There I will hear much more of your experiences. He tapped Yuri on his shoulder, who still embraced Svetlana firmly, took up Astrid's luggage and walked to the exit.

'Welcome, ladies. I had some difficulties with Interpol since you gave Feasley time to escape. But on the other hand they did not find him before. It was you both that did the major work. Where is Peasley's zwikker or what's left of it?'

To his stupefaction Astrid caught her bag and emptied the contents on his desk. Professor Bolotnikov couldn't resist laughing. What a woman, this Astrid. Emptied just from her bag the components of the most dangerous apparatus of the world. So to see it was thoroughly damaged.

Yuri, surprised too, caught some remains. 'Do you know that you could have been seriously wounded? The holder of the gyroscope is, as it happens, completely vacuum and at its destruction it implodes as if it is an explosion. The pieces do fly through each other in all directions.

'Oh, was that the reason of the boom which scared us so much. Astrid had beaten the zwikker with at chair. We had no axe available.'

'You've been very lucky for that. The chair has caught the round flying debris! The system which we developed to destroy zwikkers relies on this principle. And it destroys at the same time the user if he is in the neighbourhood. The implosion or explosion is still larger when the zwikker works. The inside gyroscope becomes a flying circle saw. I should have informed you of that, sorry. But who thinks you would destroy a zwikker with a chair? Anyway I take along the components for investigation.'

'I would like to hear Astrid's story again from the beginning,' spoke Professor Bolotnikov. 'Svetlana can complete it if she thinks that you skip something. Take your time, we want the entire tale.'

'Well, to start with...' Astrid spoke approximately three quarters of an hour. Svetlana interrupted seldomly. Jenke and Yuri listened fascinated.

'...and Peasley seemed to be under the influence of that zwikker. He was almost so far, I had that feeling that he wanted to use the zwikker more and more, and to apply it for everything he was against and "compensate" people with his own spectrum. Did you know that people would react this way? A kind of obsession?'

'No,' answered Professor Bolotnikov. 'But I can understand that people will react this way, especially if they have the zwikker for themselves. For this reason we must destroy all zwikkers outside the regular UN system. There is probably no other solution and your story confirms this. Fortunately the remainders of the only "wild" zwikker in the world lie here on my desk.'

There was however one person wobbling uneasily.

'What's the matter, Jenke?' asked Astrid, knowing her husband. He had something on his mind. 'You mean that zwikker which, how's his name again, has kept back?'

Professor Bolotnikov shot high up from his chair. 'What you are saying there. Is there another "wild" zwikker? What have you all done what I don't know.' He became suddenly very strict and Astrid recognised another side of him as a President.

'That's a long tale, Victor. Don't become angry, there is nothing wrong. It means that Ron Sheily, I know his name again, took home reserve components of a zwikker. Ben Corward of the same laboratory and Jenke have each a vital component so that Sheily cannot compose it together. They suspected Peasley having private interests since nothing had been in the press that the US was involved in constructing zwikkers. But as officials they could not undertake any action against Peasley, nor inform President Smith. That would have been immediately discovered by Peasley. They thought idealistically enough that they could perhaps intervene later if desired.'

'It becomes time to take the affairs more seriously. Is this all, Astrid, or do you have still more surprises?'

'No, Victor, not that I know. But you should not punish Corward and Sheily. They handled out of duty towards humanity. And Jenke should certainly not be punished. He has foreseen many things and prevented, although he does not get the appreciation which he deserves. Oh, yes, there are zwikker construction plans with Vice-President De Beaufort.'

'He destroyed those. I knew that already,' spoke Professor Bolotnikov. 'I think I've heard enough. There will be much to do for me and Yuri. Your task has now expired and do me a favour, Astrid, take some rest. And Mr Holthus, the question with your two idealist colleagues, I will handle, don't take any action. Do not take contact with them. You will send, however, the component of the zwikker which you have, to Yuri. We ensure the rest. Corward and Sheily will not be harmed. And now, I've had enough for this day.'

Professor Bolotnikov gave Jenke a hand, embraced Astrid and Svetlana and requested Yuri to stay.

Outside Astrid and Jenke looked at Svetlana. 'Would this be really the end of the adventure? Come Svetlana we will bring you home. Jenke and I will then decide what we will do.'

They took a taxi and said Svetlana farewell in front of her flat.

'That was a marvellous surprise when I saw you at the airport. Who warned you?'

'Professor Bolotnikov himself. I immediately caught a plane to Moscow. Your friend is taken care of the children. Come now, our plane starts in two hours. At home is also nice, I think.'

'Not such haste.... I still want to discuss with Svetlana and Yuri a number of items. My idea is that the zwikker works addictive and that the people here should know that. And, to remain practical, I must still settle up. Svetlana and I got money for travel - and accommodation expenses, but I haven't had a salary. They haven't thought about that yet. Therefore I think that we must visit today or tomorrow the zwikker centre.'

'If you think that addiction is so important, why nobody noticed that so far?' reacted Jenke.

'It surprised me too. Perhaps, because the zwikker has so far always been used by a team. But Peasley had it for himself alone. Obviously a team becomes less influenced. But people should know about it, otherwise crazy things will happen.'

'You have become wiser. I wonder whether you would miss the tension of this type when you are again at home. What do you think yourself?'

'We'll talk about that later. I would like to know whether I could speak with the Secretary-General of the UN, but that might perhaps be too difficult to arrange.'

'Oof, you really think this is necessary!' exclaimed Jenke. 'It can also be a response because you were so much involved with the zwikker. Did you already arrange the visit to the centre with Yuri?'

'Yes, tomorrow morning. Afterwards we can return home what will be delicious. Those hotels and magmobiles are perhaps nice, but, now yes, your know it all. You have maintained the garden well?'

Jenke laughed, embraced his wife and together they left for their hotel.

17

People search for company; and then they quarrel! Why?

New York

In the following weeks there was a busy consultation between the Heads of Government and the UN. The Peasley story had been brought in the publicity. The role, which had played Svetlana and Astrid, was kept secret. Peasley's photograph had been distributed worldwide, but no security service had found him. Since there was more to do than running behind Peasley, the research with the zwikker had been reduced, whereas the attention by the press had weakened.

This was also due to a serious inter-state conflict in central Africa. The reason seemed economic, but the deeper context was religious. Because there was in both states no clear separation between state and religion, a banal economic conflict had caused a war.

Dolores sat behind her desk and consulted with several staff members. 'Why has this conflict not been foreseen and prohibited?' she asked to the head consultant dr. Winfred Gomern.

'Mrs. Guerrero, we had only indications of a small economic conflict and nothing indicated to an armed conflict. Since the early 21st century few religious wars have been fought and most potential conflicts we've always been able to prevent. But it seems as if some resistant religious virus has infected these people. Hate and religion have always been the strongest motives for war.

'What do you suggest we must do?'

'We've analysed the situation, but it will be very difficult to intervene military. First there are in each country religious minorities, what makes intervention twice as dangerous, and it will last certainly a month before we could act.'

'And? What still more you have analysed?'

'One of my staff suggests the application of the zwikker. We can save thousands of lives. The question is, however, how will these soldiers be after the treatment? Anyway better strange than dead.'

'And how can you do it technically?'

'We propose to intervene from a high-flying plane with a turbo zwikker which can very rapidly measure and compensate.'

'The application of the zwikker has yet not been permitted other than for medical or technological aims. You know that. I understand from your suggestion that you ask for an exception and on very short notice. Anyway the UN should solve this conflict. We have to ensure peace and human rights, also for people as individuals. I would gladly see your plan ready by tonight.'

Dolores stood up and the consultants left her office. Then she called her secretary and requested her to call the commission "Article 1000". The UN had appointed this committee since the last far-east conflicts in order to handle urgent cases. She found that the situation in Africa legalised this method.

Central Africa

Although "Article 1000" provided the possibility of action within an hour, including discussion, it lasted still a day before a team was underway. Two new turbo zwikkers with heavy sensors had flown over from Moscow to Cairo, where they were overloaded in a special plane made available by the US. This plane was equipped with an electronic radar protection system. Professor Sven Larson and Yuri Kaspadov were in charge.

'Do you think that the zwikker can accomplish something?' asked the pilot to Yuri, when the plane flew over the Sahara in Southern direction. 'We can cover an area of 30 km on an altitude of 10 km. The spectra will show irregularities, because in this area people with several consciousness fight with each other.'

'I've taken that into account already.'

'How much time is there between measuring and compensation?'

'Oh, not much, minimum one second.'

'In one second the plane has moved 139 meters if we fly with a speed of 500 km/hour. Can you couple the zwikkers in such a way that one measures and the other compensates?'

'That's possible, but there's still the same difference of one second between measurement and compensation. As a matter of fact that one second is only valid for a small part of the spectrum. It becomes ten seconds for a complete spectrum.'

'At 10 seconds we cover 1.4 kilometres. That means that the measurement sensor and the compensation sensor must stand under an angle in order to cover the same ground surface. This angle is larger if we fly low and smaller on larger altitude.'

Yuri nodded. He commissioned his technicians to install the sensors under the required angles and asked them to carry out a number of tests.

During the flight over the Sahara the pilot, called Stan Johnson, explained how to use their ejector seats in case of emergency. Each passenger had his own one.

After two hours they had arrived above the area of the fighting parties. Several smoke clouds were visible. Stan received a radio warning that he was in a prohibited area and that he had to identify himself. He replied he was a UN plane for making observations.

'You must leave immediately in northern direction,' was the answer, 'otherwise you will be shot down!'

The pilot did not answer and continued in the same southern direction. He only started the electronic protection system which made him invisible for the radar. Arriving outside the fight area he descended to 10 km altitude and turned in a large arc to western direction. He programmed his board computer for a flight back and forwards in 10 km broad strips. The plane reacted and started at low speed of 500 km/ hour its tracks.

'Yuri, in three minutes you can start the zwikkers! I stay at 10 km altitude,' called the pilot.

Yuri put up his thump and started the two zwikkers. The technicians looked down at the battle field.

'Now!' they heard yelling Stan.

With great tension Yuri and the technicians looked at the displays of the two zwikkers. Nothing happened. 'Are there no people in this area?' asked Yuri. 'Do you see you below human activities?' he called to his technicians.

'Yes, smoke, villages and movements in the fields,' was the answer.

'But why don't we measure anything?' asked Sven. 'During the flight to here everything worked fine.'

'Something disturbs our zwikkers, that's clear. Can it be the defence system of the plane?' He realised himself that without the defence system they would fly back and forth as a defenceless insect over the fighting party parties.

'Stan, turn off your defence system for a moment. It seems to disturb the zwikkers,' called Yuri.

'Do you know what you are asking? You want to bring us down before we have started?'

'I come with you, there must a solution, otherwise we've come for nothing.' Yuri ran to the cockpit and sat down behind the pilot. 'Can you turn on and of the system rapidly?'

'Not really. It hasn't been designed as a flash light. And who knows whether the defence system will not break down after twenty times off and on?'

'Then we must modify our tactic. Stan, we will throw ourselves in the hottest of the battle. We must take more drastic measures. You warn us when you turn the defence system off and on again. Can you "see " rockets being fired at us?'

'Yes, but with 500 km/hour I cannot avoid them. I need at least half a second to turn the defence system on and that can be too late. Well, I will turn the plane now and we fly first to the place where we see most of the smoke. Hold on boys, it can become exiting' He threw the plane in a sharp turning and flew in northern direction to the area where thick smoke clouds hided the ground from view.

'Sven, we compensate only the total spectrum, everything you measure. What kind of result that gives is still a riddle. That's for later care.'

The radio cracked when Stan turned off the defence system. 'Are you still there? You'll be shot down,' he heard.

'Quick, Yuri, this is your chance, I don't know whether we will get a second chance!!'

Yuri looked at to the screen and saw a confused picture. 'Compensate fellows, fast! Everything, leave nothing uncompensated...! And save the spectrum, perhaps we must pass by again!'

The technicians did what he asked and Yuri saw the same spectrum appearing on the screen of the compensation zwikker and then its negative picture which disappeared suddenly the next second.'

'We're attacked, boys, go to your seat and attach you firmly,' called Stan loudly. The plane made a sharp turn, the engines shrieked and they took off by an angle of almost 60°. Stan had restarted the defence system.

Yuri saw three rockets passing by. Stan hadn't reacted a second too soon. 'Stay on boys! We're not safe yet. Some rockets can after one passage look through the defence system, an American invention,' he shouted. 'I'll dive now and if they come behind me I will keep the plane just above the trees on the ground!'

Yuri felt decreasing his weight and during a moment he saw a pencil floating around. They dove with the acceleration of the gravitation. Approximately hundred meters above the ground Stan redressed the plane horizontally and continued at approximately thirty meters above the hillocks. He made a turning to see if there were still rockets behind. And indeed he saw three spots which dived from the sky on them.

'We're lost boys. Our defence system has been seen through. I accelerate now as fast as possible so these rockets have to make still one turning before kissing us. That gives us time to leave the plane. On my signal you press the red button beside your seat. But do that first with the two seats with the zwikkers, so we can save them as well!'

The tension in the plane was heavy. Everyone was deeply pressed in his say, while Stan had accelerated the plane. 'Now...!! ....!' shouted Stan.

Yuri pressed the buttons of the seats with both zwikkers and saw them disappear. At the same time the seats with Sven and the technicians disappeared through the wall of the plane. Wind howled in the plane and he was almost blown from his chair. The safety belts cut in his shoulders. The engines of the plane had stopped and the plain descended steep down. In his ear-phone he heard Stan shouting: 'Yuri...!! Press the red button....!! Yuri....!'

With all power Yuri leaned against the storm wind aside and could hardly reach the red button. He tried again and pressed it..... and knew nothing more!

He came round again hanging with h s seat in a tree and heard explosions. 'I'm still alive,' he thought, 'and I have no pain, but my head spins. I can move my legs.' He twisted around, to see what was aside of him. At a kilometre distance was a large fire. 'Certainly, our plane,' he thought. There were none of his colleagues around. Not far from him lay a seat with a parachute.

Yuri removed his seat belts and drew the pieces of the parachute to himself. The seat was firmly anchored in the tree. With the knife which had been fixed at the seat, he cut the parachute in bands, tied them together and let himself fall on the ground. He was still dizzy and ran staggering to the other seat. It contained a zwikker. Yuri removed the zwikker and fixed it with the belts. Then he swung it on his back and stumbled difficultly to the closest hillock top. There arrived he saw a remarkable picture. Soldiers ran through each other, apparently not knowing what they should do. They had left behind their weapors, in spite of the fact that from far they were fired upon.

'These fellows must be the compensated,' thought Yuri, 'and those poor souls are fired by a not yet compensated enemy.'

He crawled to a part of the hillock top which was covered with heavy stones. He took the zwikker from his back and planted it on a flat stone. He aimed the sensor at the "enemy" in the distance and started the zwikker. To his joy he heard it producing the high tones which increased in height and then stopped.

'It's still functioning. The batteries are still O.K. Fortunately I had loaded them, in case of lack of current in the plane,' he spoke in himself.

Yuri had great pain to keep the sensor quiet and aim it at the "enemy". His hands trembled. He took a couple stones to support the sensor and pressed the button "measure". After ten seconds he saw the same confused picture appearing on the screen as measured in the plane. Only a certain peak was more on the right side.

'Just storing it together in the memory, and then make the compensation pattern.'

Five seconds later Yuri transmitted the compensation pattern in the direction of the "enemy" and repeated it a number of times by changing the direction. With interest he looked over the hillock top if he could detect something. Shooting continued some minutes and stopped then suddenly. In the area of the "enemy" a number of people appeared who stood up. They looked at each other and walked away in several directions. Also in the direction of the earlier "compensated", but they let them pass without any problems.

'Oof,' said Yuri. 'I've succeeded. Now in search for Sven, Stan and the technicians. Those cannot be far. He walked back to the seat in which the zwikker had been assembled and set up the ear-phone. 'Stan, Sven, can hear you me?' he called.

He heard some cracking and then: 'Is that you Yuri? Our seat batteries are almost empty because we tried to reach you for a long time. How are you?'

' O.K., except headache. Where are you?'

'We're all in a kind of bunker. We landed exactly in the battle area, but to our luck the soldiers were too astonished to take us under fire. We found shelter in a dugout.'

'Where is that? Can you specify where?'

'In the battle field, although we don't hear any more deflagrations.'

'Yes, that's correct. I have compensated them with the zwikker which had landed nearby. You can come out. Has Stan a flare or something, then I can see where you are.'

Yuri peered to the former battle field and saw suddenly shooting high a Very light. It hanged above a hillock at approximately one kilometre distance.

'Yes.., I see you where you are, stay there, I'm coming. It will take time, I've difficulty in walking and that zwikker is rather heavy.'

'Okido,' he heard weakly, then some cracking and no more.

During his excursion over the hillock back he encountered regularly people in uniform. They greeted him kindly, but were passing him without further problems. 'People without subconsciousness appear to become kind,' thought Yuri. His sense for research came up again and he would propose to Sven to examine a group of compensated straight away.

After almost three quarters of an hour he saw suddenly Stan waving. Little later the complete group descended towards him. The technicians caught the zwikker from his back and Stan and Sven gave him an arm.

'How are you, Yuri?' spoke Sven. 'We were awfully worried about you. In the meantime you've done excellent work. It was a hell here. Until the explosions suddenly weakened and stopped. Come, lie down here and drink somewhat. There is a stock of spirits in the bunker.

Yuri lied down gratefully. His head rustled. He had to fight for not sinking in unconsciousness. 'I don't remind anything from the time when I pressed the red button until I came round in a tree,' he thought. He wanted to ask Stan some questions, but his eyes closed and he knew no more.

'He must have received a heavy slap on his head,' said Sven. 'I don't see a swelling, because his head was protected by his helmet. It wouldn't surprise me if he has a heavy concussion. We must be very careful in transporting him. Stan, do you have a proposal?'

'I think you and me should go to the nearest headquarter of one of the fighting parties. Their equipment might still function and we can couple it to our pulse telephone. Thus we can reach the UN or Moscow. Those haven't heard of us since two hours. They have only detected we've used the zwikker twice.

The others stay here and look after Yuri. Who of you has knowledge of first aid?'

One of the technicians, called Arnie Zweiterbach, stood up. 'I'll take care of him. I've done that before. We will collect some weapons in case there are still some none-compensated soldiers around. Moreover, we've still the zwikker.'

Stan and Sven climbed from the bunker and started in the direction where Yuri had seen the last group of people. Evening was approaching and they should walk fast to bridge the distance of about three kilometres. There was still a lot of smoke but no more fires. It was remarkably quiet.

Yaounde

Yuri felt its head spin as moving in a whirlpool. His head bumped against something gentle. He tried to open eyes with large effort, but he couldn't succeed. 'First rest again,' he thought, 'and then start again.'

'Yuri......, Yuri,' he heard softly in the distance. 'Yuri, wake up, I'm here, Svetlana.'

Yuri tried once more and his eyelids started to tremble.

'He reacts, you see that. He tries to give us a signal,' he heard Svetlana saying far away.

In himself he answered: 'I hear you, continue I'm coming.....' With all his strength he moved his hand, and suddenly he felt that it was caught by a small cold hand. He realised it was Svetlana. Softly he pinched and got an answer. 'I feel you Yuri. Do you hear me? If that's the case, grip twice with your hand.'

Yuri pinched twice and there came a smile on his face.

'Yuri, can follow our conversation?'

He replied with two grips with his hand. He felt something wet on his cheek and heard Svetlana crying. 'Oh, Yuri, you were so far away. You want that I stay with you and talk?' Two grips followed. Svetlana looked at the people around Yuri's bed. They were Sven and Stan, a doctor and a nurse.

The doctor nodded. 'You can stay here as long as you like, he seems to wrestle to come out of his coma and you can help him thereby. I see that he has become quieter and his pulsation is almost normal. I think he is sleeping now. You should rest, otherwise you'll get ill. You look terribly pale. We will put an extra bed next to him.'

Svetlana, Stan and Sven left the hospital room and on the terrace they looked out over the splendid tropical surroundings of Yaounde. The temperature was approximately 28° C and the humidity degree was not too high.

'Svetlana listen, we must go further. We can't stay any longer,' said Sven. 'We must report in New York. But we leave behind with you Amie Zweiterbach. He is able to explain Yuri what has happened between his blackout and today.'

'I understand. You've been a great support. I've particularly appreciated that you called me directly to come over. I will bring Yuri back to life, even when I must sit a year at his bed. What do you think he had? The doctor was not very communicative.'

'He must have hit something when he left the plane. Obviously a complication has appeared when he managed to find us with the zwikker on his back. It must be small, because in the hospital there was only a small shade on the scanner. For certainty they keep him sleeping. For this reason he was not able to wake up since you arrived here early this morning. The sleep cure lasts three days. According to the doctor he should wake up tomorrow morning.'

'So he's not in coma? Really not?'

Stan caught her hand. 'Have some trust. He will manage to recover. I think you should rest now. I see they brought an extra bed. You must be dead tired after that long travel from Moscow.'

'Thank you, Stan, I wish you a good trip and let us know the reactions in New York.' Svetlana kissed them both on the cheek and returned to Yuri's room. Yuri lay breathing quietly.

The nose probe had been removed. Svetlana caught her luggage and changed and washed herself in the bathroom. She fetched her pyjama and lie down in the bed beside Yuri so she could touch Yuri. Softly she pressed a kiss on his cheek, drew a sheet over herself and fell almost immediately asleep.

The next morning she woke up with a shock. Beside her she saw Yuri sitting straight in bed. He had breakfast on his lap.

'Oh, darling, you are better and I slept.' She started crying.

'Don't cry dearest. I woke up this morning, it was still twilight and I saw you sleeping beside me. You looked so sweet that I wouldn't wake you up for anything. A nurse brought me very quietly my breakfast. Just give me a kiss. A tear-stained is also good.'

Svetlana dried her tears and laughed. 'You're still alive. I was so frightened when I saw you yesterday. Entirely pale with a probe in your nose. You didn't move and breathed difficultly. Do you remember that I've spoken to you? Your gripped my hand.'

'I dreamed you were close by but that's all I remember. But what has happened? The last thing I remember is that they carried me with the zwikker in a bunker.'

'You went unconscious and they brought you with a heli to this hospital. That was three days ago. Stan and Sven warned Moscow and New York and Professor Bolotnikov has me informed. I've been so terribly frightened. The only thing we knew was that you had been shot down. The automatic signals from the black box of the crashed plane were caught in Moscow. It lasted half a day before Stan and Sven got contact with New York and we heard that you were all saved. Oh Yuri, what I've been anxious about you. What would I have done without you? And now I lie simply beside you. I will dress myself rapidly, because some people may come in. I must look terribly.'

Svetlana hopped from her bed and entered the bathroom. A doctor came in the room.

'Ah, Mr Kaspadov, I've heard that you awoke this morning. I see that your pulsation and blood pressure are normal. And how do you feel yourself?'

'The old, doctor, or rather the young. And no more headache, a though I rather not shake my head. So don't ask me to say no. It's like there's something loose in my head. But further good. Can I stand up?'

'Not yet. We should not underestimate a concussion. Tomorrow, perhaps. And then only shortly. We will move you soon to the terrace, so you have some diversion. You can receive, however, visitors. That technician of you waits in the corridor. Shall I let him in?'

'Yes, do that. Or rather, let he go to the terrace, and Svetlana and I will come without delay.'

'You're already instructing, I hear. That's excellent. But mind, don't be too reckless. I will give your fiancée instructions about that.'

The doctor left the room and just afterwards Svetlana returned. 'Was that your personal doctor? I heard something concerning instructions.'

'You must keep me calm. Give me a kiss, that helps enormously.'

Svetlana bent to him. She still trembled.

'Can you push me to the terrace? Arnie, the technician is there and I want to speak with him. When I know everything I will become even calmer, you see.'

Svetlana disconnected the brakes of the wheels, opened the doors and drove the bed to the terrace. Arnie leaned to the balustrade and welcomed them.

'Hay, Yuri, nice to see you alive. I just spoke with the doctor. Nice man.'

'Tell me, what has happened since you brought me in that bunker?'

'To start with our crash. We landed in the battlefield and hided in a bunker lying under fire of one of the "enemies". We had no zwikker so we couldn't execute any consciousness compensation. Fortunately you did the right thing before you came to us. After you arrived, Stan and Sven walked though the battle field where lay a lot of dead and wounded soldiers. The others ran somewhat foolishly around, but after some time they started to help the wounded. It did not match which uniform they wore. Stan and Sven came during in the night in a tent which was lighted. There were some people with gold on their collar and who asked what the matter was. They knew they belonged to an army and had been involved in fighting. They only no longer knew why they were fighting and had no intention to continue. They were very kind and completely open. Stan could use the radio equipment and he made fast contact with New York. They got the instruction to stay where they were and to wait for aid. That lasted up to the next morning. There a complete army of heli's with UN staff landed close by. Some had been armed heavily, others were Red Cross doctors. Most of them went directly to the battle field. Stan and Sven got a Red Cross heli to return to their bunker. With that one we've been picked up and transported to Yaounde.

'Did you take along the two zwikkers?' asked Yuri.

'Yes, we did. When Stan and Sven left, we have gone in search for the other zwikker. We found it, or rather what was left of it.'

'Where are they now?' asked Yuri.

'Yours has been taken along by Sven to New York. The other is still here. It's seriously damaged. A grenade or rocket must be exploded nearby. But we have found all parts.'

'And what happened next? I mean with the soldiers?'

'Nothing really. They became people without complexes. They had become without exception personages. When later people from other regions showed up and gave them commands, they did not react. They wanted to go home. They simply walked away. However, they helped the wounded. During the night in the bunker and the transport to Yaoundé you've been always unconscious. We feared for a coma, but we haven't applied artificial breathing. In the hospital you have gone directly through a scanner, but you know the result perhaps already?'

'Yes, Svetlana told me. So people become kind and unconstrained as if they are entirely compensated? Interesting! Did they also forget what they had done?'

'No, I asked some soldiers what they had done. They knew exactly what they had done and why. They had been very angry with the others. But suddenly that anger had gone and with that the reason to fight. Also they didn't want to shoot back when the others still continued.'

Yuri wanted to know more details, but Arnie was unable to tell what Stan had discussed with New York. However, Svetlana could tell him that in the UN and in the world press a large upheaval had arisen because of the obstinate action of the UN Secretary-General.

'But that's not your care,' interrupted Svetlana. 'You've so far recovered and you seem to have your ideas already quite again on a row. Do you know what we will do further...? Nothing...! I will order tea and then will we enjoy this without more discussions. Moreover the view is monumental and can we spend the complete day here. If you, Arnie, would be so kind to go to the city for further news, we will appreciate that. But take your time!'

The morning passed quietly, Arnie went to the city and Svetlana and Yuri discussed future plans when returned home.

'Perhaps we can marry here, would you like that, Svetlana?'
'You think that...?'

'Yes, naturally. What would be nicer than to stay here longer as man and wife? I cannot go to work. And in Moscow it is cold. Arnie and that doctor can be witnesses.'

'What do you thinks your mother would say?'

'She will understand. She's much too glad that I'm still alive. She can join us by means of Internet.'

Svetlana leaped and cuddled Yuri so hard that he exclaimed: 'Ho, ho, I'm still a patient...!'

The doctor, who entered the terrace, intervened: 'Please leave my patient unscathed!'

'Oh, we want to marry here doctor. Is that possible?'

'Of course is that possible. We're not difficult here. Moreover, it's your decision and thereby the authorities should adapt themselves. And as I've well understood, Mr Kaspadov is a zwikker expert, so you can handle everything.'

'Oh, you've heard that?'

'It's impossible not to know. You are on the screen everywhere. As a hero. You've saved thousands of people. But to return to the marriage, it is even possible this afternoon, if you like. But on one condition. The patient should not be exhausted the first days. Can you promise that?' Svetlana blushed and nodded. Yuri grinned only.

When the doctor was gone they looked at each other and burst both in a delivering laughter.

It was exactly the moment when Arnie returned. 'What's the matter?'

'Oh, nothing, but do you have a nice costume with you?' asked Svetlana.

'Me..., me? Why..?'

'Because you're a witness this afternoon of our wedding. Nice, isn't it?'

Arnie's mouth fell open and it lasted some time before this news penetrated to him.

'Just go to that doctor. He can lend you something. And can you ensure an Internet connection with Moscow? We like our family to attend.'

Arnie's mouth closed with a bang. 'Of all….. I'm tumbling from one stupefaction into another. Just yesterday he lies unconscious and then he will marry. Would you.....?'

He wanted to continue, but Svetlana put her hand on his mouth and conducted him softly away from the terrace. 'Would you please do what I asked? Be a good boy.'

The Internet connection was installed and Yuri and Svetlana had a long conversation with Yuri's mother. She approved everything, after first having a good cry by seeing her son. Yuri asked if she would contact other family and Professor Bolotnikov because the doctor had permitted only one phone call. She promised that to do. Furthermore Yuri told her that the marriage would take at four o'clock local time and she could follow everything on the screen. She should not be able to speak, because the connection would only be in one direction.

After lunch Yuri went asleep and Svetlana drove with Arnie to the city. She didn't want to wear a nurse uniform, the only available at the hospital. In the city she hoped to find what she liked and even her own flowers.

'There are certainly there many weddings in Yaoundé,' she thought at seeing all those bride dresses. She selected a beautiful one and returned with the box under her arm to the hospital. She had only half an hour left to change, thereby helped by a number of nurses. The large news of the wedding of their famous patient had put the hospital on stalks. In the meantime flowers had been brought and the large reception room had been nicely decorated.

The hospital director came for Svetlana at exactly four o'clock. Yuri's doctor allowed him to dress and he was placed in a wheel-chair. Someone pushed Yuri to the large room where an army nurses held flowers above him. After pushing him to the podium, the nurses started to sing and dance in their own traditional manner. A priest came in and stopped in front of Yuri. Yuri saw Arnie behind himself, but no Svetlana. In the corner the Internet camera had been placed. He could see it was functioning. He waved to the camera and wanted to ask the priest some questions when Svetlana came in. She was looking splendidly and a shot of emotion went through him.

'I'm not staying in this wheel-chair. That they cannot force me to,' Yuri thought. 'I'll wait until Svetlana is beside me, and then I will stand up.'

Svetlana walked at the arm of the hospital director to the podium. A couple black little bride girls held the sleep. In her hand she had a bouquet of tropical flowers.

'Svetlana, help me to stand up and push that wheel-chair aside,' whispered he softly to her when Svetlana bowed to him.

He saw that his doctor disagreed, but there was nothing he could do.

What Yuri and Svetlana did not know, was that the ceremony was transmitted over the complete world. When Professor Bolotnikov got the phone call of Yuri's mother he had immediately contacted the Russian Broadcast Company. He wanted to make people aware of the historical action which had performed Yuri. The transmission had been preceded by pictures of the last battle and interviews with compensated soldiers. Pictures of corps and wounded had been shown, as well as the shot down plane and Yuri in unconscious condition.

Fortunately Yuri and Svetlana did not know. The priest pronounced the Orthodox Christian formulas and gave subsequent a short catholic benediction. Their yes was answered by singing of the nurses, who danced around the podium. They replaced Yuri in his wheel-chair and brought him back with Svetlana to his room. There stood a connected threetel and the bride couple could receive for a short moment the congratulations of their family and friend being together at a reception in Moscow.

'How many people there are, and there is your mother,' said Svetlana. 'How they all made it?' To the camera she called: 'Thanks, darlings, we will celebrate it again at home. When Yuri is fully recovered we will come home.

The doctor who had walked along with them stopped the threetel. 'I'll leave you a moment alone, but in ten minutes Yuri must rest again and I trust you look after him. I congratulate and hope you will live long and happy together. I could not have foreseen this yesterday. But Russians seem to have hard skulls. That was your luck, Mr Kaspadov.'

Left alone Yuri stood up and embraced his bride for the first time. 'Svetlana..., I hope we will be very happy together.'

'Sure dearest, and I will take care for that. But now you become pale again. Come, I will take off your clothes and help you in bed. I stay there until you sleep. I'm not taking off my bride dress. n a couple of hours I will awake you and we will take together our wedding diner. There won't be any guests. Just we two. In fact, I like it this way, entirely together.'

Yuri slept almost immediately when he felt the sheets pushed in his back. The hypnotics of the previous days had still their effect. Svetlana stayed with him until his breathing became quiet. She slipped gently to the terrace. Outside the sun set in a red glow.

'How I'm happy,' she thought. 'I'll remain sitting here as long as possible. Naturally I must thank of course the nurses and also the priest. But this moment I will enjoy alone. He lives, he's not dead. I will never forget this moment.'

New York

The world did not forget it either. The retransmission had done more than any zwikker could have done. Yuri, Svetlana and the city of Yaounde had become famous in one blow. Joy to live and naivety had defeated hatred.

Dolores had followed the transmission at home. What those two people had done she couldn't have reached in months in the UN. She had felt the last time a large uncertainty. There was no way of return. The UN would be able to defeat worldwide conflicts. And what to do about those ultra religious who turn themselves into terrorists. How to reach them? They could hardly compensate the population of whole nations into unconstrained children.

She spoke long with Jorge, but they could not find a solution. She hoped that more malignant people would cross her path to help her.

18

How to criticize a leader who under the cloak of his ideas attacks another.

New York

'You both have observed that your own people and those of the adversary demonstrated in principle that they did not want a war. When de-doctrinated, we call it ce-compensated, they have no reason to fight. In other words, you have incited them.'

The two Heads of State had, after a long hesitation, agreed to come at the expense of the UN to New York. They had been accompanied by a general and religious leader.

`You call yourselves leaders,' spoke Dolores further. `Do you realise, that we don't live anymore in the twentieth century when millions of people were killed in the most terrible way, only for reasons of political ideas, faith and passions?'

'Mrs, I believe you don't understand us. Not the government caused the conflict, but the people themselves. And the government is their exponent. In my country religious cores had been formed who got support from the population. Going on the street and yelling slogans. The same happened in the neighbouring state, only those had an opposed opinion. We as Heads of the Government have really tried to prevent a conflict. But it resulted in spontaneously small armed conflicts and this led inevitably to the large conflict where the armies have been used. If we had not defended ourselves, this would have conducted us to downfall and anarchy. Say yourself, where would we have ended?'

A religious leader interrupted and said: 'Through your action part of our population has lost their faith. Do you realise that?'

'Gentlemen, do you really believe this is worth thousands of dead? But have you forgotten that a government is nominated for the people and not the other way around? What sense has it to adapt to a religion if one cannot live together with the neighbour. Most religions preach peace, but let me tell you that the UN will never permit a war justified on the basis of principles. And I will tell you why. In principle you assassinate your own people since they must sacrifice themselves for their faith. And you yourself continue to observe that from a distance. Soldiers have been trained to execute orders. But what about their own critical mind? The application of the zwikker has shown that by the compensation the soldiers regained their capacity to reflect and think clearly. He doesn't let command himself any longer to actions which he really doesn't want!'

A silence fell. Dolores's guests looked at her unbelievingly.

'Dare you, as you sit here at the table, to follow a treatment with the zwikker? Your complexes are then removed and you can think uninhibited and clearly. Perhaps a true faith as meant in the Bible or Koran repossesses you. I've arranged a zwikker studio and before you undergo the treatment, you can decide if afterwards you want to be brought back into the old situation.'

'But..., but, you haven't the right to ask that,' called the religious leader who had spoken before.

'Oh no...!' spoke Dolores angrily. 'The UN can do therapeutic research. It will be carried out by the group of Professor Larson. You can be brought back afterwards in the old situation. If at least, you have still that desire.'

'I think I accept,' said the religious leader from the other party. 'I request Professor Larson to make a video recording, so that I can decide later what I want. That seems reasonable to me.'

'You can get it as you like, I guarantee. The others can watch from the screen here in this office.'

Dolores brought him to the adjacent office where Professor Larson placed him in a chair which stood in the middle of the studio and started the teleconnection. The test went as expected. After ten minutes the religious leader returned in Dolores's office.

'If I understand well I'm now a compensated person. I must say it feels refreshingly. I've no ulterior motives concerning any problems and I would like to remain feeling this way.'

'And how you feel with respect to your faith?' asked one of the Heads of Government.

'Oh, marvellously. I can now see that the writings must be explained differently. I must return to the source of the human contact with God. I think that I will be a better religious leader than before since I will focus on the core of the truths.'

'Do you want to be brought back in your old situation,' asked Professor Larson.

'No, certainly not. I feel myself much improved. I remember everything, but that oppressive feeling has gone. I would say that I'm at last delivered of the original sin.'

'Are the others interested in a treatment?'

None of them reacted. Dolores repeated the question, but she got no response. 'Gentlemen, our meeting has finished herewith. But please remember, the UN will not tolerate any new war operations. Your arms are already confiscated and you won't receive any financial compensation. You've only reached a real victory when you become wise enough to lead your own state.'

'I don't need to tell you that you will have to justify yourselves for the General Assembly of the UN. A call to this end you will receive soon. You had luck that at the shooting down of a UN plane nobody was killed.'

To the not-compensated priest she said: 'I think you should go with your compensated colleague in conclave. This could perhaps solve the problems between your both countries. It is no shame to discuss. You can serve your people to a large degree.'

She accompanied her guests to the corridor. It was remarkable, that these experienced leaders had been impressed by Dolores's attitude. As reprehended children they left the office of the Secretary-General. Only the compensated priest had a smile on his face.

'You're some woman,' spoke Jorge, when she had told the complete tale of the visit. 'You've touched them in their primary-school-mistress complex. They feel if you have put them in the corner. But what means the recall for the General Assembly?'

'You know that since the early 21st century no more war has been allowed by any of the member states. If it nevertheless happens, the warring parties must appear before the UN. In former days there was a kind of Security Council, but it has been abolished, as you know, due to the fact that too many nations wanted a fixed seat. Now warring party parties must justify themselves before the General Assembly and explain that all peaceful resources had been exhausted and that their operation was unfortunately nothing else then self-defence.'

'But if these Heads of Government don't do that?'

'Then they are struck off as member of the UN and automatically they are boycotted in all areas of economy, transport, communication etc. If such a Head of Government puts one foot outside his country, he is arrested and summoned before the International Court of Justice. But, in general, this is an exception.'

'What will happen with all those compensated soldiers in Africa?'

'Their behaviour is studied. We look how they react and how long they will remain uninhibited persons. A group of psychologists which gained experience in Uzbekistan is acquired to go there.'

The next evening turned out differently than Dolores had expected. Instead of being satisfied, she was increasingly worried.

'Dolores, I fear you've come under the influence of the zwikker,' said Jorge. 'You seem always searching for possibilities of using the zwikker. And the way you speak about it is unusual. Wasn't it that woman from Strasbourg, who you sent to Brazil, said that Peasley had become under the spell of the zwikker?'

'Perhaps you're right, Jorge. I'm dreaming of that thing. In my dreams I want to brainwash whole populations with the zwikker, in order to make it a better world. What would that be? I nevertheless haven't been frequently in contact with the zwikker. Only three times. Two times at demonstrations in the Central Hospital and one time in my office.'

'When have those dreams started?'

'I think after that second demonstration. And they became more clearly after last week. You think it's serious?'

'Yes, I do think so. And you must take immediately action with those people who work in New York and Moscow with the zwikker. They must be entirely influenced. They probably don't know it by themselves, since they work permanently in a group. Has somebody left from these groups?'

'I don't think so. No, now you say it, we are astonished about their devotion. Nothing is too much for them.'

'Then it's your task to start a procedure which prevents that people to become completely under the spell of that apparatus. You're fortunately not yet influenced far enough, I hope at least?' Jorge caught Dolores by her hands and felt them trembling. He held them for considerable time until Dolores slowly relaxed.

'Fortunately, my internal zwikker still works. But please, come never again in the neighbourhood of that apparatus. You promise that?'

Dolores nodded. Her eyes became clear again. 'What do you propose?'

'You can't approach those people who work with the zwikker. They will do everything to protect themselves. Remember they are addicted and will show all the signs of drug addicts. What do you think of those two ladies who found Peasley? They know the problem and they can be trusted. They haven't been in contact with working zwikkers.'

'Not a bad idea.'

Strasbourg

Thus it happened that Astrid was interrupted by the telephone during the preparation of the diner. On the screen she recognised the image of the Secretary-General of the UN. 'Mrs Holthus do you have a moment for me? Confidentially? Please test your code, then we can talk undisturbed.'

'Oh, rather, you want to speak with me? Let me first turn off my potatoes, otherwise they might burn, I'm afraid.' She walked away and returned a moment later. 'I've pressed the code, Mrs or can I say Dolores? Thus we you know all. Victor also calls you Dolores.'

'Which Victor you mean, could you tell me that first?'

'Professor Bolotnikov of course and I'm allowed to call him Victor.'

'Well, it be Dolores, as a matter of fact you did already. That talks also easier. I need you for a difficult job. It concerns the addiction of the user of a zwikker. You noticed already something with Peasley, when you met him in Uruguay.'

'And how can I help you?'

'First let me know how you think about the equipment.'

'The apparatus has disappeared quite from my mind when I returned home to Jenke and my children.'

'Are you able to come over for a couple of days in New York? I invite also Svetlana Kaspadov with as argument that the UN want to give you both a medal of honour. The real reason is different, since we three must find a solution before it is late. Can you come?'

'Of course I'll come. Can Jenke come too?'

'In fact that would be correct for the medal. But rather not. Then I must also invite Yuri Kaspadov and he does nothing else than work with zwikkers. You understand?'

'Yes. I will be, however, honest to Jenke, he sees immediately through each excuse. When should I come?'

'The day after tomorrow, at ten o'clock in my office, is that possible?'

'Yes, Dolores, I'll be there.'

Astrid still asked: 'How do you think of keeping away Yuri?'

'Don't worry, I'll handle that with your Victor. A ticket will be provided tomorrow.'

Astrid waited until her image zoomed away and jumped up. 'Joopy, how nice. I'll meet again Svetlana and perhaps we find a solution. Jenke must stay home. He has already been brave earlier.'

That evening Jenke got a badly prepared meal in front of him. Salt was lacking at the vegetables and the potatoes were cooked to pulp. He looked at his plate and then to Astrid. 'What's with you, why are you hopping around? And this food hasn't any taste.'

'Do you know who rang up this afternoon? Guess once?'

'Your mother?'

'Cold, try again?'

'Svetlana?'

'Warmer. No, Dolores Rodriguez Guerrero in own person. On that display device there. She looks very smart. How do you find that?'

'And what about it? You have to go again? I hope not.'

'Dolores, I can say Dolores. I've to be the day after tomorrow in New York to receive a UN medal together with Svetlana. For services provided. Isn't that nice? Furthermore there is still something about which she wants to speak with us.' She nodded with her head to the children who had followed the conversation with interest.

'Woopy, Mum, then you are famous! You will come on the threetel and you must say a word of thanks. Can we help you with that? Such as "I haven't at all deserved that medal, since everything was ample chance and thus...."'

'Rascals, that I will absolutely not say. Of course I deserve that medal and I will tell them that too.' Astrid laughed to her children. Together they philosophised further and forgot the potatoes.

'And what was that other thing?' asked Jenke when they were alone.

Astrid told about the addiction to the zwikker and the worries of the Secretary-General.

'And she prefers I shouldn't come along?'

'Yes, under no circumstance, otherwise Yuri should come too. He must be already addicted to the zwikker and will do anything to disturb any solution.'

'Yes, I get it. I shall have to invent a good reason to stay here. One may ask at my work why I haven't accompanied you. I think that sickness can be an excuse. Each other reason sounds suspicious. But do you have already ideas for a solution?'

'I've thought about that all afternoon. The addiction can possibly be measured by the zwikker. And if so, it can be compensated.'

'But would the addicted fellows agree to undergo thus treatment?'

'I suppose not. Do you have a suggestion?'

'Perhaps. You must have someone who is able to trace the addiction in the spectrum. Then each zwikker must be equipped so that it compensates the "addiction" peak automatically.'

'And you think that something like that is possible?'

'It is probably more complicated, but you can start with this principle don't you think?'

They heard the children above clatter. They were washing themselves.

'I think that you're right,' said Astrid after a while. 'I will make list of what I must take along. Do I have to carry evening dress at the distribution of that medal? Svetlana can just take one from her cupboard, I suppose. Can I buy tomorrow a new one?'

New York

'You are Mrs Astrid Holthus?' she asked. 'I'm Soraja, secretary of the Director General. "Please follow me to the office of the Secretary-General.' Svetlana was already present and both frends embraced each other warmly. Then Astrid greeted the Secretary-General who invited them to take a seat.

'Ladies, I appreciate it particularly that you could come on so short notice. You, Mrs Kaspadov, are not yet informed on the real reason.'

'Not again tracking? I thought we were here to receive a medal of honour? Not that is necessary really, because I haven't done that much.'

'No, it's not tracking, but something for which I selected you both. The zwikker seems to work addictive on the user and this problem can become only examined by people who are not yet addicted themselves, but who know the conditions. For this reason you had to come alone, since your spouse works already for a long time with the zwikker.'

'Addiction...? People get under the spell of that apparatus...? Now I understand why I'm more and more worried. Yuri can behave strange and he dreams at night aloud about his work. He has been therefore also addicted...?'

'I suspect so,' said Dolores. 'My husband noticed it too with me. I wanted to use the zwikker for rising problems. He was the second who suspected addiction. You Astrid was the first.'

'You both must help me. I cannot play this along the official canals without strong proofs. And the fellows in the Central Hospital and Moscow are certainly against. Astrid, have you already reflected on what I said by the telephone?'

'Yes, Dolores,' upon which Svetlana hopped on her chair when she heard that they were already on first-name terms with each other. 'We must try in a malignant manner and by means of Sven Larson whether the addiction is visible on their personal zwikker spectrum. In other words we must erase addiction by means of the zwikker. I sense we must start at the source. It is of later care why people become addicted. The zwikker irradiates probably some type radiation which causes this addiction. But first these persons must become normal again. Perhaps it is even necessary to compensate them entirely.'

'How do you want to tackle this? We can't handle the apparatus by ourselves and if we work with the zwikker we become addicted too! And we aren't even allowed to work with it.'

'We must be malignant and apply the rules to break through the system. Can we have an own office in this building? And if possible, a permanent connection with Professor Larson's laboratory? I think that we must work by means of him. But then in such a way that he does not see through all details. That will be difficult, because drug addicts are, in whatever form, very suspicious.'

'I'll give you my own reception room. For the first weeks I don't need it. There's a telephone connection which is protected, so that nothing can leak out. I'll give you a technician who can be trusted. Is there something else you want?'

'Can we consult you if that becomes necessary?'

'At any moment of the day. This is the number of Soraja, she can always find me.' Dolores stood up and led Astrid and Svetlana to the door.

'It's not yet clear to me what we must do,' started Svetlana in the corridor. 'Do you think we might succeed?'

'If you want to help Yuri, we have no other choice,' was the sober answer. 'Let us visit the reception room. When time comes council.'

When arriving they found Soraja who presented them to the technician, Pierre Rabeaud. He smiled widely to them. 'I'm entirely at your service, *Mesdammes*, it's me a large pleasure to cooperate with you. You say but what I must do.'

'Provisionally nothing, Pierre. We can say Pierre.., yes? I'm Astrid and this nice lady is called Svetlana. Can you install here Internet? And a computer which can be coupled to the telephone? If possible before lunch?'

'*Naturellement, Madame*, or Astrid. That's no problem. I will prepare everything in my workshop and install it here.'

Pierre left and Astrid and Svetlana stayed behind with Soraja.

'To what extent are you informed on our mission?' asked Astrid.

'I'm always informed on everything what concerns the Secretary-General. Don't worry. If you need something just ask me.'

'What I need is a signed letter that we are allowed to cooperate with the group in the Central Hospital. The letter should specify that we want to examine the impact of the zwikker on the user. And also that we have the right to work with their zwikkers by remote control. Can you arrange that?'

'No problem. There are some drinks in that refrigerator and Pierre can order a lunch, which you can consume here.'

'That's not a bad idea.'

Soraja had hardly closed the door behind herself when Svetlana said: 'You are going to use the zwikker yourself...! Isn't that dangerous...? Suppose you do something incorrect and those poor fellows become entirely disorientated.'

'That danger is present. I must get them to the point that they cooperate voluntarily. With your collaboration I will tell them a tale about Yuri. Or do you have something better?'

Svetlana thought long. 'Let us go through all facets of the problem once more. If we are discovered during our attempt, it is possible that they encapsulate themselves, Yuri included. And I don't want to loose him. I've not married a zwikker, but a man who I love. Let us be particularly careful.'

'Perhaps we should first detoxinate one of them. For example Sven. He's the head of the group and knows most. Once he's cured of his addiction, he can help us further.'

'But how? You say yourself that someone who is addicted protects himself. Moreover, they are the last ones to admit a problem of "addiction". They just don't know.'

'Then we must handle the zwikker at distance.'

'Yes, perhaps. How you think of getting Sven up to that point?'

'I should work on his feeling for scientific research. We can try that in any case.'

'And can you handle the zwikker?'

Astrid shook her head and wanted to say more when Pierre Rabeaud came in with a cart full of material. He asked the people who helped him to place the equipment on the large table and thanked them for the effort.

'This the best equipment we have and that's a computer which we can couple to the telephone. It is O.K. that I install them?'

'Go ahead, Pierre, we will drink something in the canteen. We are back in an hour when lunch is served. You're invited, of course.'

'Am I lucky, I need only three quarters of an hour to build everything together.'

During lunch Astrid and Svetlana recapitulated their plan whether there were weak spots.

'Well, Svetlana, will you ring the Central Hospital. According to Soraja, Professor Larson is present.

She was smoothly put through and saw appearing Professor Larson on the screen.

'Hay, that's nice, to see you Svetlana. How is Yuri? From where you ring? I see the number is from New York.'

'Hallo Sven, yes I'm here in New York. Yuri is rather O.K. He has recovered from his injuries and restarted his work. I'm here with Astrid Holthus, you know her name. We're at the UN. We'll receive tomorrow morning a medal of honour for provided services. Isn't that nice?'

'Oh, I would be glad to come, but unfortunately that's impossible. We're in the middle of a busy programme. Perhaps I can make later some time free to meet you. Is Yuri with you?'

'No, Yuri stayed in Moscow. But don't hang up, Astrid want to speak to you.'

'Must I? I'm very busy.'

'Yes please, it's on behalf of Yuri.'

'Professor Larson, or may I say Sven, as I you know from the tales, please call me Astrid.'

'My pleasure, Astrid. What's the matter?'

'Briefly said, Yuri has not recovered completely. Svetlana and I are here asking for help. Yuri has side-effects which he himself and the specialists do not understand. One suspects that it still has to do with his accident in Africa. It's too long to explain you everything. We got authorisation from the Secretary-General to work at distance with the zwikker and examine Yuri in Moscow. You wonder perhaps why this is asked to us. That comes because we know everything of the zwikker, except to work with it in practice. Furthermore we cannot do it the centre in Moscow. It is necessary to carry it out at distance. We're here to learn and you must help us thereby.'

'But for that I have under no circumstances enough time. Our group stands to do measures in a Catholic church and in a Mosque. It has lasted very long before we received authorisation and the measures will start this afternoon. We absolutely don't have time. Perhaps I can permit you tomorrow half an hour, but not more.'

Astrid saw the complete plan fall through. She realised that Sven had become already a fanatic zwikker addict.

'That's splendid, therefore till tomorrow, and at what time?'

'Ten o'clock precisely,' said Sven, looking at his note book.

'Do you have this afternoon perhaps a technician in your laboratory, so that we at least gain some time by learning to work with a zwikker at distance?'

'That's possible, indeed. One zwikker remains in the laboratory and there is always one technician available. You know what we do, I leave behind Boris Klemarov, who is most experienced. Is that O.K.?'

'Splendid and thank you very much. Is the research you carry out giving nice results?'

'Particularly interesting. But youngsters, I have no more time now, Boris will ring you soon. Tomorrow you must explain exactly what's wrong with Yuri.' He hanged up and zoomed away.

'Cor,' said Svetlana, 'that was within an ace or we've been with empty hands. Now we can at least start.'

'Pierre,' said Astrid, 'there's work to do. You've heard we must learn this afternoon as much as possible from Boris Klemarov. I think, Svetlana you should speak Russian with him.'

'We must get Boris on our side. After that we can catch Sven and subsequent the complete group. But it's not so far yet. You said Pierre, everything is ready?'

'Yes, except the zwikker, everything works.'

'You know,' whispered Svetlana, 'you haven't shown Sven our authorisation.'

'It's good he didn't ask, otherwise he might have smelled danger.'

As promised, Boris Klemarov appeared some minutes later on the screen when the telephone buzzed. 'How I can help you, ladies? Professor Larson demanded me to instruct you on the use of the zwikker at distance. That's also new for me, but I am entirely at your service.'

'Are you alone, Boris?' asked Astrid.

'Yes, all the others have gone. Why do you ask that?'

'If we will practise with the zwikker, we want gladly a volunteer for measures. And now you are obviously that volunteer.'

'If you think so. But I don't want that something happens with me, you understand. How much experience do you have with the zwikker?'

'Nothing what concerns its working. That's exactly what we must learn and then still at distance. But what I, however, know is that we must store your own spectrum safely in case we make an error. We have here a technician, Pierre, who can help. Furthermore I am assisted by a countrywoman of you. She's called Svetlana. Here she is.'

Astrid pushed Svetlana forward and for some time the conversation went further in Russian where they only distinguished the words such as Pierre, Astrid and Yuri. Then Svetlana gave the word to Astrid.

'What you've discussed has escaped me, but I propose that you first will achieve the connections with Pierre. Send us on our screen the guide so that Svetlana and I can study that in the meantime.'

With an enthusiasm, technicians own, Pierre and Boris started. Astrid followed their conversation and read at the same time the guide. It appeared soon that the handling of the zwikker was rather simple. It was only difficult to identify parts of a spectrum. It lasted approximately an hour before the technicians were ready.

'Ladies, we are ready,' spoke Boris. 'I'll give now a small introduction, to start with showing a number of spectra which we have put away of examined persons. You can take these over in your computer and study them later once more quietly. But one thing, I have joined a programme that the spectrum erases after use. They're protected against copying. To guarantee the privacy, you understand?'

An interesting demonstration was shown, which Astrid, Svetlana and Pierre followed with large attention. Especially the part concerning compensation.

'That's it,' finished Boris satisfied. 'Now you know everything and I suggest you run through my presentation once more. Other questions?'

'Yes, this afternoon we want to carry out a real measurement with you. We want to measure your spectrum and discuss with you the details. Is that possible?'

'Let us say at three o'clock. I'll prepare everything in advance with Pierre. O.K.?' Boris grinned.

'Excellent, and we will do our home work and we call you prompt at three o'clock. Many thanks for what you have learned us.'

'For the moment he reacts quite normal,' whispered Astrid. Louder she said: 'Pierre, we've to study everything once more again.'

Pierre helped them through the copied presentation of Boris. After approximately an hour they knew everything and took then a pause. Pierre left and promised to be back three o'clock.

'Now comes our hour zero. We should not fail, because it's the point whether we will get a second chance. If Boris sees through it before we've compensated him, we've lost everything. He will protect himself and inform all the others of the group. He might even ring Yuri and then we are real far from home.'

'Oh, Astr d, I hope we will succeed. I don't want to loose Yuri. You know exactly what you must do?'

'I hope so. As soon as we have Boris' consciousness spectrum on our screen, we must make its compensation spectrum and return that immediately. Pierre must have the compensation programme already ready. Or do you think that we must do it ourselves. Suppose Pierre says something to Boris before we can compensate. Then Boris has enough time to turn off his zwikker.'

'I think that we must program the computer during the absence of Pierre. Svetlana you're ready to press the compensation button as soon as the compensation spectrum appears on our screen. Then Boris still lies under the sensor.'

They repeated everything once more before Pierre returned. It was almost three o'clock.

'Pierre, would you please connect us to Boris, would you? Svetlana and I to have everything repeated in all details, at least we think so. You don't have to handle the computer. That does Svetlana and I ensure the conversation. Go quietly in that seat in the corner after you have achieved the connection.

Pierre made the connection and Boris appeared on the screen.

'I've everything ready, ladies. The zwikker is already warmed up. Tell me what I should do. I'm your man.' He talked as if it concerned an ordinary job.

'Do you have an interesting consciousness spectrum, Boris?' Astrid smiled to him.

'Rather, I've somewhere a giant complex, but it is your task to trace it. I've told you how to find it.'

'Yes, yes your presentation hadn't been wasted to us. And I bet with you that we will succeed in one turn. On what we bet?'

'If you succeed, I invite you both for diner tonight. I know an excellent restaurant. But don't tell it to my wife in Moscow. She's trusting me, but you mustn't force it.'

'Boris, Boris...., and what must we do when we don't succeed? Not that there's a chance.'

'The same, but then you pay, O.K.? I am a big eater, so better try to succeed.'

'Well, let us start. Would you please lie down on that bank above which hangs the sensor? And remain very quiet afterwards. You can project the spectrum above your head so we can discuss it at the same time.'

'Is that necessary? You've nevertheless my spectrum on your screen?'

'Yes, but I want a second spectrum after I've asked you a couple of questions. I want know whether the spectrum changes in the most recent part. Is that O.K.?'

'If you insist. According to the fellows here something like that doesn't happen.'

Boris lied down on the bank and Astrid gave Svetlana, who was out of view of Boris, a signal. She started the measuring and both on the ceiling above Boris and on the screen appeared the total consciousness spectrum of Boris. What Svetlana noticed immediately was a strange peak. They had not much time to think about this because they typed directly "compensate". In the meantime Astrid talked with Boris and asked the questions which she had prepared. She made the questions as long as possible to gain time for Svetlana to complete the compensation.

'Do you have the answer already on my first question, Boris?'

'Yes, but I find that question rather indiscreet. One has its privacy, isn't it?'

'But that's exactly the point we should discuss. Have you yes or no piddled in bed as a little boy? I see a crazy peak in the left half of your youth spectrum.' Astrid kept her face under control, not to burst in laughter about the large Russian on the bank.

'I don't answer that question. Soor you will ask me still about my intimate life before I was married. And then the complete world will know.'

'Not the complete world, only we. Soon we'll have forgotten all, really. And by the way you can let us forget it with your zwikker, dc you? From the corner of her eye she saw that the compensation spectrum on the display was completed. She gave Svetlana a signal, who reacted immediately.

'Astrid,' said Boris, 'if you had paid better attention you knew that the zwikker cannot erase a memory, only consciousness.. ... Hay, what happens there...! I feel myself light in my head...! What are you doing..?'

Boris still lay on the bank and seized his head. He rolled back and forth and Astrid was afraid he would tumble from the bank. A moment Astrid looked at Svetlana who indicated with her arm high up that the compensation process was still busy. Only just before Boris started to sit up Svetlana lowered her arm, indicating that the compensation had been completed. Pierre had also leaped up, not understanding what had happened. He had only looked at the display of the telephone, so he missed what Svetlana had done.

'Pierre!! Don't touch anything!!' shouted Astrid, 'keep your hands off! We know exactly what we've done!' Pierre withdrew startled. He had not expected such an outburst. There was obviously more at hand then he was told.

Boris who had also heard what Astrid said to Pierre, sat on the edge of the bank and called: 'What has happened? I see no more spectrum on the ceiling and I feel light in my head. Not unpleasant by the way.'

'You've won your bet, but at the same time lost it,' reacted Astrid. 'And for understanding this riddle, I propose you to turn off the zwikker and come to us. Anyway you're our guest tonight for diner.'

'O.K. if you want that, I'll come. I'm curious to know what has happened. With a defect zwikker I can't work for Professor Larson.' They saw on the telephone screen that Boris stopped the zwikker, took his coat and cut off the connection.

'And you,' turning to Pierre, 'you've behaved exemplarily.'

'But I've done nothing...?'

'That's what I mean. If you had, our complete plan could have failed.' Astrid tapped Pierre on his shoulder. 'Look, boy, Svetlana and I are here for a special task and for that we had to compensate at least one person of the group of the Central Hospital. And we succeeded. We will discuss this later with Boris. He doesn't know yet, therefore keep your mouth when he arrives and don't be astonished on the result.'

'You're really something. In the future I'll take better care to cooperate with women.'

Aiming at Svetlana, Astrid said: 'My congratulations. The first step has been taken. Now we will treat Boris further. That must also succeed, especially if you explain everything in your mother language that you've done that to save Yuri. That will touch him. We must no longer play hide and seek with him. He will react normally, I think.'

It took Boris only half an hour to arrive. Svetlana and Astrid were slightly overwhelmed by his enormous body. On the screen it had seemed normal, but standing beside Pierre, he seemed a giant. Astrid grinned at the idea that this large man had piddled in his bed.

Boris embraced first Svetlana, who had greeted him in Russian. Astrid got an enormous hand.

'We appreciate enormously that you've come straight away. We need your help, more than we showed initially. We've entirely compensated you, do you realise that?'

'Oh..., is that it! That's the reason I feel myself different...! Much lighter, although you wouldn't say that, ha.., ha.., ha...,' Boris laughed thundering.

'But you don't find it terrible?' asked Svetlana careful.

'Certainly not, girl,' and he lifted her from the ground. 'I feel myself delightfully, eagerly to dance. Perhaps we could do that tonight.'

'Do you realise what we did and why?' asked Astrid.

'And..., and... would you be so kind to put me down again?' said Svetlana.

'Sure, girl, you're so light that I forgot I still carried you. And what your question concerns, I certainly realise what you've done, but I don't care at all. But please explain. That you've compensated me is not insurmountable. I can repair it easily. My spectrum has been saved, isn't it?'

'Yes, we have it here in our computer. But there's no hurry for that.'

'No problem. Now, why you've done it? It was therefore no mistake. If so, I had won my bet. In fact now also, because this was not our agreement.'

'Your bet you've won anyway, Boris,' Astrid smiled at him, but becoming serious: 'Sit down quietly and we tell you the complete story. Why we're here and what we must achieve.'

Astrid spoke of the addictive functioning of the zwikker on its user and that they had got the request from the Secretary-General to find a solution.

'But why Mrs Guerrero hasn't simply commissioned this to us?'

'Because you all are addicted to the zwikker. Only I hope this is no longer the case with you. Is that right?'

'You said all of us are addicted? How do you mean?'

Astrid gave the word to Svetlana who explained all details to Boris in Russian.

'Perhaps you're right, I realise now. And if I am honest, my ideas twisted more and more around that thing and I dreamed of it sometimes. Now I'm not feeling anything. For my part I never work with it anymore.'

'Then we have at least cured one person of your group. But our task is to cure everyone before it is too late. Would you like to cooperate with us?'

'How?'

'First we must know why the zwikker users become addicted. Then the zwikkers should be modified so this can no longer happen. As a technician you think that something like that is possible?'

'I don't know. If those zwikkers should be modified, it can only be done in Moscow. But why the user is addicted, we must examine them all together. Our group here and that in Moscow.'

'It's not that easy, Boris. When all drug addicts will work together, isn't that a time bomb for the world. So we must first decompensate all users here in New York and in Moscow.'

Boris stood up and paced heavily up and down in the room. 'Only from a distance I think, as you did already. But nevertheless people must measure with that thing in order to repair it. If we have Professor Larson on our hand, we can manage perhaps the rest also.'

'And when do you think we can do that?' asked Svetlana.

'Tomorrow. I will play the same game with him as you've done with me. Control test of the zwikker by remote control. O.K., I'll try to arrange that with your equipment and Pierre can help me thereby.'

'And we look from a distance,' said Astrid, 'because it's better when a Professor Larson doesn't see us.'

They recapitulated again what to do, and Boris went home with the positive promise not to go to the laboratory. Pierre brought Svetlana to her hotel. Astrid stayed for informing Mrs Guerrero. Afterwards she held a long conversation with Jenke.

Astrid was not dissatisfied about the obtained result. How she had to proceed, she didn't know yet, but perhaps other experts could assist. She thought of Ben Corward and Ron Sheily. However, for the moment the fewer people involved, the better. Dolores had pressed her to keep secrecy, and she had said nothing to Jenke. She had only spoken about the ceremony of the distribution of the medal which would take place tomorrow. Dolores had said that the ceremony would be simple, but that a small word was expected from her.

'Well,' thought Astrid, 'let us first dine together. But I'm not going to dance with that bear. I'll leave that to Svetlana.'

Dolores reflected about Astrid's report. They had succeeded to cure one person from his addiction complex. This proved that zwikker addicts could be healed. She looked at her watch. It was seven o'clock in the afternoon and actually she wanted to go home. But she knew also that Victor in Moscow was waiting for her bulletin. She pressed the intercom and saw that Soraja was still present.

'Can I do something for you?' she asked.

'Yes, can you get Professor Bolotnikov on the phone, in spite of fact it's night in Moscow.

It lasted only two minutes before Dolores saw Victor in pyjamas appearing on her telephone display device. 'Have you achieved success?' coming straight to the point. He had waited all night for her bulletin.

'Yes. Your two ladies have cured the technician, Boris Klemarov, a compatriot of you, from his zwikker addiction. They lured him with a trick and it succeeced. This proves that zwikker users become addicted. You have to take in Moscow measures, otherwise things run out of control. The ladies have shown the possibility of using a zwikker by remote control and compensate one's consciousness spectrum. If you do that, the person is cured from his addiction. At least, as long as he does not work again with the zwikker.'

'My compliments. This is a great relieve. What do you recommend?'

'Your people of the zwikker centre must be treated. But I doubt whether this can be done as easily as with Boris.'

'You said that the zwikker can be handled by remote control?'

'Yes, we have here at the UN equipment which is linked by computer with a zwikker- of the Central Hospital. Thus Svetlana and Astrid have compensated Boris who was in his laboratory. Their plan is now to compensate first Professor Larson.'

'Again by means of a trick?'

'Yes, and let us hope this goes well, too.'

'Don't hope too much on your luck. Suppose Larson smells danger. He is an experienced zwikker user and he is certainly heavier addicted than Boris.'

'Could we only keep them provisionally away from the zwikker, but that's the difficulty. Larson and his group are in town with two zwikkers.'

'Then I know what I will do. I'll clear the zwikker centre in Moscow directly. I'll summon all employees at the Ministry and send subsequently Nicolai Yesin with a security group to the centre and lock up all zwikkers. I will address the employees and notify them that it concerns a security test. They will be brought somewhere outside Moscow. Afterwards we will draw up a compensation plan.'

'But if those fellows protest?'

'Than I'm sorry for them. We cannot let any addiction continue. It's not an easy measure and even Yuri can become dangerous. But I must accept that. Tomorrow I take contact with you and perhaps we have a plan to solve the matter.' Victor broke off. He was obviously in a hurry to prepare the actions.

Dolores remained silently in her office. Should she do the same with the group of Larson? But the zwikkers were at three different places, she understood? One in a catholic church, a second in a Mosque and third was still in the Central Hospital. Since she had understood that Larson would come tomorrow in his laboratory, she chose for the plan which Astrid had presented. Without a compensated Larson it would be impossible to cure the others.

19

Drug addicts will protect their addiction at all cost. The arguments are always the same.

Moscow

The closure of the centre in Moscow hadn't proceeded very smoothly. Some employees refused to come to the Ministry. A number of them fled into the city. Furthermore, the security service could not open the strong-room where a number of zwikkers were stored. The result was that they acted rather roughly. The centre was sealed off and the electricity was cut off.

When Victor overlooked the excited people, he felt a strange threat mounting from them. 'These people aren't prepared to listen,' he thought. From the platform he tried to find Yuri, but how much he looked, he didn't see him. 'Nicolai, these people won't swallow any tale. Remove them as soon as possible and lock them up and let nobody escape. Their common addiction can lead to dangerous actions. Lead them one by one to the corridor, handcuff them if necessary.'

The gloomy scene which took place was more depressing than Victor had suspected. People were violently resisting and could only be overpowered by force and lead away by additionally called up police. 'Who ever would have thought this,' he sighed.

Nicolai joined him later at the platform: 'There's only lacking one person, Yuri Kaspadov. He was not in the centre and we haven't found him at home neither in the city. The whole Moscow police force is after him but so far no track.'

'That's terrible. Yuri is a good friend and now this. As the inventor of the zwikker he's probably more sensitive to the zwikker than others. He has certainly noticed that something was going on. Otherwise we would have found him, don't you think? Redouble your search, and take care that nothing happens to him. Remember, he's a national hero.'

Yesin left after he had given necessary commands, while Victor went to a quiet place and caught his card-phone. He dialled the private number of Dolores. It was still eleven o'clock at night in New York.

In view came Jorge. 'Hay, Victor. You'll want Dolores? Wait a second, I'll call her. Dolores..! Victor on the line...!'

Dolores came up rapidly and spoke: 'Everything went well?'

'No, we locked up all employees after much resistance. Fortunately nobody is really hurt. But what is more serious, Yuri was not among them. We think that he has sensed it and went into hiding. The research centre has been sealed off hermetically. We must rapidly start to compensate the employees. As soon as that has happened, there is no reason to keep them locked up. Have you already a plan?'

'Oh yes. After I you rang you lately, I've asked Astrid for ideas. Her answer was that we should compensate by remote control. That is possible from New York. She proposed that Boris should assist thereby. As soon as we are ready in New York and I'll keep my fingers crossed, Boris goes to Moscow and installs somewhere one zwikker. Than we treat them by remote control from the UN building. You place each employee of the centre in a drawer and shift him under the sensor of this zwikker. We'll do the rest, understood? Of course we can build also a remote control installation in Moscow, but that takes too much time. Moreover we are sure that our's is functioning.'

'Perhaps not a bad idea. You give me some hope again. If only Yuri would reappear. I'm worried about him.'

'May be Svetlana should return?'

'Perhaps, but not yet. It can also work contra productive. Please don't tell her about Yuri. Also not to Astrid. Tell them all goes well here. See you soon.'

'That's become exiting,' said Jorge who had listened. Those Russians have tackled the matter quite hard. I'll reflect on what Yuri could have done in this situation. Perhaps I'm getting tonight a bright idea.'

New York

The trap which Boris had set up for Larson succeeded above expectation. The suggestion that his consciousness spectrum contained some unexplainable peaks and could only be measured with "remote control", had sufficiently aroused his scientific curiosity. Thus Professor Larson lied, just like Boris the previous day, on the bank and Astrid and Svetlana could execute the same operations as before. Boris conducted the conversation with Professor Larson and did that masterly. Boris talked about all kinds of technical tricks that Professor Larson almost fell asleep.

The response was the same as yesterday. Professor Larson seized also his head when the original consciousness spectrum disappeared from the ceiling.

What afterwards followed was easy. They asked him to turn off the zwikker and come for a meeting with Mrs Guerrero. There he received a complete overview and reacted rather astonished. Furthermore he was equally unprejudiced as Boris. He agreed entirely with the setup and thanked Astrid, Boris and Svetlana.

'You saved me from worse, isn't it? I'll help you further treating the others. That shouldn't be too difficult, indeed. They have followed me so far, so why not again. But one thing they will find strange, that I'm no longer in the laboratory. You, Mrs Holthus must help me to invent an excuse. I've understood you're terrible good in eye washing.'

'Not better than anyone else, I suppose. The best thing is to approach reality. That's my experience. Only my spouse I can't frame. But these fellows don't know me. Well, I'll help you. Boris and you should work from here, and your employees will let their spectrum deviations measure. We say no word about compensation. Subsequently they must come here to discuss those deviations.'

'Not bad, that discussion we will have anyhow. We must know where that addiction has its peak in the spectrum.'

Dolores, who had followed everything, congratulated the scientist with his recovery and approved the plan. 'As soon as we are ready with your group, we will help the people in Moscow. We have no to time lose!'

These thoughts damped the joy of their two successes and especially Svetlana looked a bit off colour. She thought of Yuri and hoped that everything was well with him. He had been rather absent minded yesterday on the telephone, exactly as if he was expected something. 'Could I only go home,' she thought. 'Perhaps the day after tomorrow. I'll ring him today; finally he must know how that medal ceremony had been this morning. That I almost forgot.'

Boris, Astrid and Larson discussed the strategy of their plan, whereas Svetlana returned to her hotel. There she tried to reach Yuri, but he was not at home. Also his mother didn't know where he was. She hadn't seen him for two days. The telephone from the centre didn't respond. Svetlana felt herself increasingly worried. It was very uncommon that she could not reach him.

She was still puzzling about Yuri when Astrid knocked at the door. 'Oh you know, Astrid, I still can't reach Yuri. He's still not at home and neither with his mother. And the telephone of the centre does not answer. What do you think is going on? The last remark of Mrs Guerrero wasn't very encouraging.'

'Svetlana, I can see that but you have just married. That boy is obviously busy and is somewhere occupied. Perhaps the complete centre is making a trip?'

'But then there should still be some guard to answers the telephone. Now I get a strange buzzing sound and no picture on the screen ..' Svetlana looked at Astrid for help.

'I think that we shouldn't panic. Yuri is involved with something and who knows he will ring later on the evening. Let us have some food. I'm hungry. Afterwards we will screen Internet for news from Moscow. Well?' Astrid tapped Svetlana on the shoulder. 'Come girl, we've always been able to solve our problems.'

She didn't give Svetlana more time and took her to the lift and the restaurant.

Moscow

Meanwhile it had not escaped by the press in Moscow that something was up at the zwikker centre. Several family members had rung the newspapers whether they knew where their family members were brought to. A number of journalists asked the Ministry what was going on, but they got zero on their request. Only after long insisting, the Ministry mentioned that a security exercise was held which would last some days. After that the press would be informed entirely.

Astrid and Svetlana found after diner this information on Internet. 'So that's the reason, why Yuri cannot ring. Such exercises are always held suddenly. Suppose everybody rings first his family, then such an exercise is worthless.'

'I hope you're right. Let us search for other individual bulletins.'

Svetlana searched further, whereas Astrid listened to her translation in English. There was nothing about the zwikker centre. One spouse of a technician who worked there complained that she hadn't been informed on the security exercise. She had to learn it from the press. She worried about her husband since he was the last time so unstable. He dreamed al the time not being able to finish his work. She insisted that her husband should have a sick leave after the exercise.

'You see,' said Svetlana, 'they're also addicted. Would that exercise be related to the addiction? I hope that Yuri will ring soon.'

Her wish was fulfilled within half an hour. She startled of the buzzing of the telephone and saw Yuri appearing on the display. He looked confused and seemed to be in a hurry.

'Svetlana, are you there? I cannot see you, because there's no display device here. You must help me. They've taken away all personnel of the centre and the centre is closed. I'm in the cellar without light. There is only a small emergency lightning. I can't use it too much, because the batteries are almost empty. Accidentally I was working in the cellar of the centre and could thus escape. I don't know what's behind all this, but I've brought one zwikker in security. I can only contact the outside world by means of an old telephone in the corridor, of which the display is lacking. Do you see me?'

'Yes darling I see you. Oh you look awful. I'll come straight away and will help you. But how can I find you?' cried Svetlana.

'I don't know. I'm somewhere under the centre and all doors are locked. There's perhaps an escape by means of the Abbey, but I don't dare to use it. The Abbey will be guarded, I suppose. I cannot give you the number of this telephone, because there's no more number on it. I won't use it again, because they might detect where I am. Come rapidly, find a way to reach me and bring some food with you. Here is nothing, only water in a toilet at the end of the corridor.'

'I'll come and will find you. Hold on Yuri...! I'm coming...!'

Yuri's image faded away. He had broken the connection.

Svetlana burst in tears. 'I'll go directly to him. I'll catch the first plane. But how can I reach him?'

'I've an idea. I'll ring Jenke. He knows that fellow who stole last year the zwikker plans. He entered the centre, although we don't know how. Jenke will help you. As soon as you are underway you can expect help of him. He will help you definitely.'

'Oh, what can I do without you? Can't you come along, you always know a solution?'

'No, I stay here. I will tell them that you're unwell and keep that appearance for a couple days. Ring me, however, always. On my pulse telephone. I'll help you pack and bring you to the airport.'

Astrid called a cab and within an hour Svetlana stood at the counter and checked in for a flight to Moscow by way of Strasbourg. The direct flight to Moscow had just left and this was the fastest connection.

'Svetlana, in a certain way it's good that the flight goes by means of Strasbourg. I will ask Jenke to meet you at the transfer desk in Strasbourg.' Astrid embraced Svetlana.

'Poor girl, she does not realize for which she comes to stand. How far will Yuri be addicted? But a woman in love can be tougher than the strongest and most malignant man.'

Astrid returned to the hotel and rang directly Jenke in Strasbourg. He came on the display extremely sleepy. 'Hey, why you wake me up? We talked this afternoon?'

'Darling, listen and start our private code.'

Jenke did what asked Astrid him and rubbed his eyes. 'What's this all about?'

'What's the name of the fellow in your department who stole in Moscow the zwikker plans?'

'Cheese, do you call me for that in the night?' From her expression, however, he made up that he better did not ask silly questions. 'Travenko, I believe, yes, I know it again, Ilja Travenkov.'

'Now, just listen and ask no more questions. It's very urgent what you must do. Svetlana will land soon in Strasbourg and takes afterwards a connection to Moscow. You must try to find out how Travenkov entered the zwikker centre. Yuri is locked up in the cellar of the centre and you must bring Svetlana to him by means of the same way as Travenkov did. Dress yourself and go immediately to Travenkov.'

'O.K. darling, but tell me nevertheless what's going on. I need at least a good argument for Travenkov, otherwise he won't tell me his secret. He's not really keen on me after what I did to him last year.'

Astrid told Jenke in short what had happened. They had to save all zwikker addicts. The only chance for Yuri was Svetlana. If she would not succeed, Yuri might become a danger for his surroundings.

Strasbourg

Jenke, who had become in the meantime completely awake, put further no more questions. 'I know what I must do. Perhaps I'll ring you tomorrow morning from the airport what we will do, otherwise Svetlana will ring you from Moscow.' He broke the connection and woke up his children Ove and Joanna.

'Youngsters, dress you, we go to Moscow with the first flight. Later on I'll tell you why.'

'Must we, pa?'

'Yes, we must and you'll become heroes when you simply do what I say. I haven't enough time to explain now. Mamma rang and she sounded rather agitated.'

He left the children dressing and searched the house number of Travenkov. Fortunately it was not on a red list and after one minute somebody took up. There was no picture. The Travenkov's had certainly disconnected this. What he heard was a sleepy voice with a heavy Russian accent which said: 'Hello, who you want to speak to?'

'Is this the house of Travenkov? I am Holthus, a colleague of him.'

'Oh, that's you, you speak with Natasha his wife. Why do you ring in middle of the night?'

'Is your husband at home? I must speak to him terribly urgent.'

'Ilja is on an official trip, Mr Holthus, but if you give the message to me I can reach him perhaps.'

'Where is he?'

'In Washington, but I can him reach there if you want.'

'Oh my God, that has no sense, because I want him to meet me in Moscow. Do you how he entered last in that zwikker centre? You were with him. I must know that tomorrow morning. We must save someone who is locked up in the cellar.' Jenke saw already his plan fall through.

'Oh, Mr Holthus, you don't need him for that. I can help you better than Ilja. I worked in the Abbey, as it happened, and I know the underground corridors which link the Abbey with the cellar of the centre. I've even still the master card.'

'Thank you, thank you….. Can you come along tomorrow morning with me to Moscow. I arrange everything for the tickets.'

'Of course, I'll be ready in an hour if you want.'

'Well, I come to you in an hour with a taxi and underway I'll tell you as much as possible about the reason of travel. I am particularly grateful to you that you want to help us.'

'Where will you stay in Moscow, Mr Holthus?'

'I've no idea. Some hotel I think. I've my two children with me. I can't let them home.'

'Perhaps I can help you. There's a small hotel in the neighbourhood of the apartment of my son-in-law. I will reserve two rooms there. My daughter can show your children Moscow if they wish. Is that something for you?'

'Marvellous, thank you. I'm very relieved that you are willing to come along and help me. I'm with you in an hour. Your address is still the same as given in the telephone guide?'

'Yes, and it is easy to find. See you soon Mr Holthus.'

Jenke rang subsequently the night number of the airport and made a reservation for Mrs Travenkova and themselves for Moscow. He prepared quickly a breakfast and filled a trunk. 'That's nice, Daddy. What a delicious surprise. Suddenly travelling and not to school. Has it to do with your travel to Washington last January?'

'Yes, but I'll explain it later. Eat your breakfast rapidly. I must also leave behind a message for my office and your school. There is no time to ask authorisation. I will ask that afterwards.'

They found Svetlana in the transit hall of the airport. She looked tired. Svetlana recognised Jenke directly and felt hope rising in her heart. It surprised her that Jenke was companied, but having lived with Astrid, it didn't astonish her any longer. Jenke presented Mrs Travenkova who put Svetlana in Russian at her ease.

The plane was not fully occupied. Natasha sat beside Svetlana and Jenke sat between Joanna and Ove. Ove was at the window. Towards midday they landed in Moscow and went straight to the Abbey.

Moscow

'Oh, Svetlana, are you there? I'm so cold. The heating has been turned off. Yuri clasped Svetlana in his arms. They stayed long together, whereas tears run on their cheeks. Yuri trembled over his complete body. He had filled up his clothes with packing materials and looked like a moon man. In the obscurity he had little space to move in order to keep warm. It was approximately five degrees above zero.

'How did you find me?'

'With the help of Natasha Travenkova, she's here. Give her a hand. Svetlana aimed her portable lamp at Natasha who had remained on the background. 'Come. We must leave immediately. I've a warm coat, a fur cap and a scarf with me, so you will look the same as all the others on the street.'

Svetlana and Natasha caught Yuri at the hand and together they ran through the corridors and ended up in the Abbey. Jenke and his two children joined them and guided them to a magmobile. There stood a police officer on the parking, but he let them pass without notice. Yuri, who walked with difficulty, shivered of the cold. In his coat and with his fur cap he resembled an old man.

'Sit in the back and don't show you too much,' said Jenke. 'The complete police force of Moscow is searching for you. What do you have there under your coat?'

'A zwikker. But I'm keeping that firmly. It's mine. Don't touch it....!!' Yuri had shouted the last words rather sharply.

'I won't Yuri, we're to help you.' He took the wheel and drove rapidly away.

'Sorry for my outburst, but I'm not really myself the last time. I haven't slept for two nights. Where are we going?'

'To a small hotel close to Natasha's daughter. There Svetlana and you can take some rest.'

'Is there threetel?' asked Yuri. 'I've something urgent to do.'

'Be calm, Yuri, calm down. I'll put you first in a warm bath and then you must eat something.' Svetlana caught Yuri's hands which were blue and trembled. He was entirely upset.

'Do you know what has happened in the centre? I heard people in the centre shouting the day before yesterday and there was some fighting. I happened to be downstairs with a zwikker and hided myself. Then it became quiet and the lights and heating went out. I found an old emergency telephone and could reach you in New York. Afterwards I've stayed in complete obscurity. Fortunately I found some packing material as isolation under my clothes. I had left my winter coat above in the laboratory. I've searched for an exit, but all doors were locked. I'm so glad you found me. What's really the matter...?'

His pale face and wild eyes did fear the most terrible.

'They have a kind of security exercise. All employees have been removed from the centre and brought somewhere else. It is Professor Bolotnikov who has ordered this exercise. I think, however, that they have missed you. Why he ordered this exercise, I don't know. For security reasons, it was said. What do you think yourself?'

'I've no idea. Perhaps there were people who wanted to steal zwikkers and Professor Bolotnikov wanted to prevent this. I had lastly more and more of those ghastly intuitions. For this reason I must do something when we are in that hotel. Suppose they would find me earlier. That would be a calamity.' Yuri caught Svetlana's hand. 'You must help me, do you?'

'Naturally, darling. That's why I came to you. They don't even know in New York that I'm here. I travelled on my maiden name. And nobody at the airport paid attention to me. What must you do so rapidly?'

'That I'll tell you soon. Are we almost there?'

'Still five minutes,' said Jenke over his shoulder.

Ove and Joanna sat quietly in the back of and examined Yuri attentively. 'Are you really the famous inventor, Mr?' asked Ove.

A tired smile slid over Yuri's face. 'That's what they say, but I haven't the feeling now, boy.'

'May I know what you've invented?' asked Joanna.

'Of course, but then you must wait until tomorrow, is that possible for you?' Yuri's smile became something broader. There was even some colour on his face.

'Hoy, Joanna, tomorrow we will zwikker,' exclaimed Ove. 'How you like that?'

Yuri's eyes started to glow. 'Ssst, boy...!! Don't use the word zwikker. The walls have ears. And I must do first something secret, you understand?'

Ove and Joanna looked scared around at the walls of the houses and observed Svetlana. She nodded. 'Everything will be all right. Perhaps Yuri can you show something beautiful tomorrow.'

They had arrived at the hotel and Yuri and Svetlana got off first. They looked around, but nowhere a suspected person could be seen. Svetlana led Yuri to his room and filled up the bath with warm water. She undressed Yuri which let himself help willingly. 'Oh, how deliciously, that warmth. I was cold until my soul. Stay with me. I must see you.'

'I'll stay with you, but I'll first look for some soap in my luggage. She went to the room and rumbled around.

'Don't touch my zwikker...!!!' sounded it suddenly from the bathroom.

'Of course not, my darling. I'm only searching for soap. Svetlana shrank from that yell. She had difficulty to master herself. Yuri became completely upset if one approached his zwikker. He must be terribly addicted, much more than then those in New York.

'What must you do, Yuri?'

'That's nothing of your business...!!!' shouted Yuri. 'Keep off my zwikker, I told you....!'

Svetlana returned to the bathroom. 'Yuri I'm here to help you, know that. I'll do everything you want or do not want.'

Yuri brushed with his hand over his face if he searched for support from far-off memories. His expression became again quieter. 'Can you look up the programmes of the threetel? There should be a direct transmission of something very important. Look rapidly, I'm leaving the bath so we can start to work. If you really want to help me, do exactly as I say and ask no more questions. Can you do that?'

'Of course, my boy. I'll do exactly as you say.'

She ran to the room, started the threetel and looked at the list of programmes for that day. With the keyword "direct transmission" she found a number of local events and one concerning the heavy winter in the northern hemisphere. There was a debate in the UN concerning the disastrous consequences of reduced oil production on the world economy.

'Which one do you mean, Yuri? There are many about our region, but only one transmission concerning a debate in the UN.'

'Yes..., that one. Please focus on that channel.'

Yuri entered the room only half dressed and looked at the screen of the threetel 'Yes, that's it. Sit down there and do not come between me and the screen. Yur designated the chair which was farthest from the threetel.

'Stay there and ask no questions. I must do this and each intervention is perilous.' Yuri's voice had a dangerous sound and Svetlana trembled of fear. She considered what she could do and reminded the action of Astrid who had destructed the zwikker of Peasley. 'Should I do that?' she thought. 'But suppose Yuri would attack me then?'

The situation in which Yuri found himself, she decided to do nothing. Yuri was dearer to her than all the people who he obviously wanted to treat with his zwikker.

As from a distance she heard Yuri mumble. 'Now start the zwikker and aim the sensor at the screen. I'll open the combined spectrum which I have collected last weeks. Then I will take the spectrum of the fellows who debate there. Just wait it is complete.' Yuri remained silent for a minute. 'Now add this bla-bla adds spectrum to the other..., O.K.' The mumbling became somewhat calmer. He sat beside the zwikker and pressed, as Svetlana could see, on the compensation button. On the screen Svetlana saw appearing the compensation spectrum.

'So, now will happen what I must do...!!!' shouted Yuri. 'Svetlana, stay there and say nothing...!! I can't allow absolute nothing...!!!'

Yuri pressed the "transmit" button and after some minutes there happened something strange on the threetel picture. Svetlana saw people at the General Assembly of the UN listening to a participant who stopped suddenly and seized his head. Dozed off delegation members became awake and stood up. A kind of confusion arose, exactly such as Svetlana had read in the press bulletins of Uruguay.

Yuri had not only erased the consciousness spectrum of all delegation members but also of all people who had followed the threetel transmission. Svetlana wondered what she should do. Yuri sat stiffened in his chair. He had closed his eyes as if he was totally gone. A nervous tick marred his face.

Svetlana became frightened. What was the matter with him? The zwikker was still aimed at the screen with the transmission button pushed in. Svetlana stood up inaudibly. Gently going to the sensor of the zwikker she twisted this away from the threetel screen to Yuri. Then she seized Yuri first gentle, but then firmly with all her force.

Yuri awoke frightened and tried to shake off Svetlana. 'Didn't I tell you not to disturb me under any condition!'

Svetlana grasped him still more firmly and said nothing. She prayed she would have the strength to restrain him. Yuri wrestled to come loose, but Svetlana held him with all the strength she had.

Then she said gently in his ear: 'Yuri, Yuri, I'm it, relax. It must, you hear.'

Yuri relaxed a moment, but struggled again to detach himself.

'Yuri, listen to me, stay calm!' Svetlana cried of exertion. She had to continue until the zwikker had done its work. She felt her head spin but sustained until she felt herself fainting and fell on the ground.

In the meantime there was a large panic in the office of Professor Bolotnikov. The satellite service had communicated the Ministry that since ten minutes a zwikker was operational in Moscow. Professor Bolotnikov received the news, briefly after Nicolai Yesin had come to his office.

'Where was that zwikker operative?' asked Professor Bolotnikov.

'Somewhere in Moscow, Professor. You know that the satellites can only determine the position within a rage of some kilometres. For that we require pocket detectors.'

'Go straight to the zwikker centre and collect a couple of these detectors. Call the heli police and search the city systematical y. We must find that zwikker as rapid as possible. Keep me each three minutes informed whether that zwikker still works and if it has moved.'

Yesin disappeared and Professor Bolotnikov rang the number of Dolores. When she came somewhat later on the display, he asked: 'We need Boris directly, because the matter runs out of hand here. Since twelve minutes a zwikker is busy in Moscow. I think that Yuri has escaped with a zwikker and he is using it. For what reason, is a riddle. The security service is going to trace him with heli's. But independent of that, send rapidly Boris to us.'

'I'll do what you're asking. I know from Larson that they will finish today the compensating of his group in the Central Hospital. Nothing can go wrong. Afterwards we send Boris Klemarov in a special plane to you. Perhaps I send Svetlana with him, although she had caught a cold and has remained in her hotel'

'Like you wish, but I need Boris as soon as possible here!'

'Calm down, Victor, everything can be handled. I must now return to the General Assembly where they have a rather important discussior. I have to be there.'

What Dolores did not know was that when she had left for responding to the call of Professor Bolotnikov, Yuri had aimed the zwikker at the threetel and had started with the compensation. That had been completed when Dolores returned. What she observed was beyond her imagination. The speaker had stopped and the delegation members walked foolishly around.

'What's the matter, Hermann?' she asked the President.

'Nothing, Dolores. The speaker said suddenly that he was saying nonsense and stopped his presentation. The delegation members did not even protest. They started to chat kindly with each other. I don't know. I've the same feeling. Problems don't exist any longer. We people must solve mutually everything in harmony. Look, there is a beautiful conversation in the corridor between people of the Northern States and that of Africa. Let us go for a tea, Dolores. I close this meeting with some appropriate words.'

President Hermann Janitsch tapped on his microphone and spoke: 'Dear delegation members. You have all understood it. We can face uninhibited the future. I close this meeting in the hope that you can handle mutually your problems.' He tapped with his hammer on the table and left with Dolores the hall.

During the tea Dolores reached rapidly to a conclusion. What had taken place, was a school example of the zwikker effect. That must have done Yuri.

'To where was the meeting transmitted, Hermann? I see that threetel groups were present.'

'The meeting was worldwide transmitted. Only in countries where it is now night, they will give a replay later today. But by means of the satellite connections all people could follow the meeting on ground.'

'Then Yuri must have compensated millions of T.V. viewers in the world.'

'Oh, is that the marvellous feeling I have. I feel myself so quiet and happy. I haven't felt myself so well in years. Yuri must a have statue.'

'That's what you say now but there are much more people who are not compensated, people such as myself. And that can be a danger for all those happy fellows. What must I do?'

'Nothing, why be so busy. There is nothing the matter. Understand that.' Hermann placed a hand on her arm. He had a cheerful smile on his face. 'By the way you're looking very sweet.'

'Oh, my God,' thought Dolores. She removed softly the hand of Hermann, stood up and went to her office.

In Moscow the population heard shrieking heli's above the city which seemed to search something. They flew first far from each other, but came gradually together and turned finally in a small circle around a certain point. They had traced the zwikker, which was still operative.

In the streets police mobiles hunted in the same direction and stopped in front of the hotel where the signal of the zwikker was strongest. This information was passed on to Professor Bolotnikov. He ordered the police to wait until he had arrived. A heli would bring him.

In the meantime and Ove, Joanna had Jenke had noticed the activity of the heli's and the police mobiles. 'Quick children, there are something going on. We will warn Svetlana and Yuri.'

When they knocked at their door there was no answer. Jenke, becoming very worried pushed his shoulder against the door. The lock succumbed and he invaded the room. There he found Yuri on the ground with Svetlana half on top of him. Both were unconscious. Jenke overlooked the scene and saw that the sensor of the zwikker had been aimed at both of them. He acted quickly, kicked the sensor in another direction and jumped over them to the zwikker. He fiddled nervously at the buttons, not knowing which was the stop button. Suddenly he saw the only light on the zwikker extinguishing. A high sound arose which weakened slowly.

'Ove, help me to carry Svetlana and Yuri on the bed! Joanna, fetch a glass of water from the bathroom to let them drink when they recover!'

Both Yuri and Svetlana looked blue pale and breathed heavily. Their hands were ice-cold and trembled.

'Massage, boys. You do the same with Yuri what I do with Svetlana. Take off their shoes and start each with an arm.'

Jenke and the children were so much concentrated in their work that they hadn't noticed that people had entered the room.

When Professor Bolotnikov landed he got the information that the zwikker signal had stopped. He run with Yesin to the lobby and asked in which room the guests Kaspadov were. The manager recognised the President and preceded him to the third floor where the new guests stayed. At the broken open door they stopped and listened to the sounds which came from inside. Professor Bolotnikov was the first who entered.

He noticed Jenke which called: 'Hey there, come here and help! These people are deeply unconscious and ice-cold. Help my children. Do not stand there, hurry up.'

Professor Bolotnikov had recognised Svetlana and Yuri. 'That's Yuri,' he thought, 'but wasn't Svetlana still in New York?'

Victor took off his coat and took over the massaging from the children. To the children he said: 'Go for help, there are people in the corridor. Call for an ambulance.'

Ove and Joanna, who found this extremely exciting, rushed from the room and bumped against Yesin. 'Call an ambulance, Sir, quick. That man who came in said that you should do that.' Nicolai took a look in the room, and hurried away.

After a half hour of massage there appeared slowly some colour on the cheeks of Svetlana. She started to tremble over her complete body.

'Svetlana,' called Jenke, 'wake up! Don't let you sink away again.' He covered her with a blanket and lifted her head. He put the goblet with water to her lips which the children had put on the small table.

Svetlana's eyelids vibrated and suddenly she opened them slightly. With a shock she awoke and gave a hard shrieking yell and looked anxiously in the distance. Then she closed her eyes again and trembled violently.

'Mr Bolotnikov, I see now it's you, these people have had a strange shock and must be rapidly treated. Can you arrange that?'

'Certainly.'

'Do it fast, because I'm afraid that something terrible has happened with them. When I came here they lay on the floor and that sensor was targeted on them. I've kicked it away and turned off the zwikker.'

'You tell me later the details. We will first help these people and then you go along with me to give further explanation. Are that your children?'

'Yes, I took them along. Astrid is still in New York.'

The instructions of the President were rapidly executed and within half an hour Yuri and Svetlana were brought to a hospital where a number of doctors were ready to help them.

Jenke had wrapped up the zwikker and followed with Ove and Joanna Professor Bolotnikov to his heli.

20

*When half of all wolves become sheep, the fences are down.
Can one find the core?*

New York

The consciousness crisis, which Yuri had caused world wide, had spread a large alarm under the Heads of States and groups of population who were not compensated. Approximately one quarter of the world population seemed to have followed the threetel emission. Governments had been disrupted because part of the ministers, members of parliament, senators and civil servants no longer realised the necessity of political actions and found the existence of their state unimportant. On the other hand these compensated were completely satisfied with the existing rules, as far as concerned public order and traffic.

'Why change any more? If something works, we don't need more politics. Aren't we all friends?'

The same also happened with the compensated religious believer. They didn't renounce their religion, but they were completely tolerant towards other conceptions and found that traditions were subordinate to the feelings which rose in them.

The not-compensated realised rapidly that these compensated had to be approached much different as had been customary. They lacked entirely obedience with respect to authorities, although they were prepared to do things when asked. They lacked also aspirations to seem more important than them were.

Within the UN the panic was still larger than in the different countries and churches. Almost all delegation members had been compensated by Yuri and only one part of the UN officials had remained free. Dolores hardly knew with who she should discuss and how she could continue to run the programme and settle current matter. For this reason she took the only obvious decision to proclaim a break of half a year and requested all delegation members to return to their country.

The President of the UN, Hermann Janitsch applauded with joy at this idea. 'There's in fact nothing important. All those items on the agenda are making live only more complicated and as you could see from the last general meeting, the delegation members found this too. There are a couple of bores who were not at the last meeting, but that's a minority.'

Dolores had looked worried at him. 'In what will this end up?' she wondered. 'How long will people remain compensated? Will a new consciousness develop automatically, or will the old consciousness return? Fortunately Jorge had not looked at that broadcast, so that he wouldn't greet her with that cheerful smile. And what about people, with a bad disposition who would exploit the compensated. How could they defend themselves against perversity?'

Another matter was, how she could use the "cheerful smiles" period for bringing more peace between populations and countries and abandon for ever terrorism and extremism. The beautifu conversations between the delegation members in the UN at the last meeting would give for that no sufficient guarantee. But with whom she must cooperate?

It was now a week after Yuri's despair action. Dolores was in her office. It was quiet, nobody rang. Only Soraja was as usually behird her desk. Would Victor ring? He had not been compensated. But Victor had his own problems with the Russian Federation. That Federation seemed to fall apart as loose sand and he hadn't enough time to rearrange his government.

'Nevertheless I'll phone him,' thought Dolores. 'I must have someone to on fall back on, and then not one of those soft-boiled eggs.'

She got the connection after ten minutes. 'I must speak with you. Do you have time?'

'It's irrelevant whether I have enough time. The world is as such in movement that I can do little than trying to solving problems step by step. What do you want to know?'

'First of all how are Yuri and Svetlana and the centre employees?'

'To start with the last ones. They are reasonably well. All have been compensated by Boris and Larson and are now satisfied people who have resumed their work in the centre. With work I mean only research with zwikkers under remote control. Furthermore the research of the anti-zwikkers is accelerated. Concerning Yuri and Svetlana I'm seriously worried. You know we found them unconscious in that hotel with the sensor targeted on them. Obviously that lasted too long. Normally people become after a compensation process quiet and happy. Yuri and Svetlana have remained one day unconscious. When they came round, it looked whether they were entirely empty. Physically there was nothing wrong, but the first days they only stared in the distance, as if their mind was elsewhere. We have laid them beside each other, so they could at least touch each other. We think that this has saved them. It was as if they returned from very far. When they recognised each other, they have held their hands. They got a music therapy and that helped. Svetlana has played for the first time on her violin.'

'How is their memory?'

'Normal, I would say.'

'They will fully recover, you think?'

'I don't know. When I visited them, Svetlana played. Her violin has got an extra dimension. I've never heard something so special. You could say she was in heaven.'

'Have they told you what has happened with them?'

'No, nothing. But something happened with them or rather inside them, that's for certain. The compensation should not have lasted any longer, I think, they would have slipped away. Died, without any physical cause. Dolores, that zwikker is much more dangerous than we've thought. We must speak about that soon and decide what we must do. But first I have to solve here many problems, too many. One of them is that a number of extremist groups, all non-compensated, are harassing the population. We can't fight them with the military, that would end up in a civil war. We will use zwikkers with remote controlled heli's and have them compensated.'

'But when these groups capture such a heli, what happens then?'

'Something very unfortunate for them. It will explode.'

'You use hard measures.'

'Less hard than shooting. If we don't act quickly, things will run out of hand. I'll take care of the Russian federation and I suggest you start with your "friend" de Beaufort of Europe and Al-Said of Africa. We should leave out the ordinary UN procedures, most of the delegates are like wet chicken. I'll phone you later this week about the details, is that O.K.?'

Strasbourg

It looked already spring when Dolores landed with a small staff of experts in Strasbourg. Julien De Beaufort met her on the airport.

'Welcome,' he said formally, and invited her to take place in the President magmobile. 'You've meddled the world quite a bit since our last meeting in New York.'

'Thank you for receiving me with these words. I see you haven't changed. You still take it evil that we continued to develop the zwikker?'

'Not personally. But wasn't I right? That thing has undermined the structure of our States because the compensated people have hardly interest in politics and authority. And isn't the State based on these principles?'

'Julien, let us leave the subject for what it is. You, Al-Said and myself must come with a good basic plan. And time presses. And you are anyway involved as a key person if you want or not. Fortunately you and Al-Said are not compensated.'

'What happened with the centres for zwikker research in New York and Moscow? I would gladly know that before we meet Al-Said.'

'Nothing to worry about. Provisionally the zwikkers will only be used by remote control. Does that satisfy you?'

'Not at all. Is that all, you haven't any more information?'

'No, you must do it with that until we meet Al-Said. Are you still in the possession of the plans?'

Julien startled. 'No, and you know that. I destroyed them. I found them too dangerous. A box of Pandorra, that's what the zwikker is, not more and not less' He spoke on a shrill tone.

'Calm down, Julien..., everything is under control.'

'I help you hoping. I am not at all reassured.'

They had arrived at the EU building, where the Egyptian delegation was already waiting.

'President Mohammed Abrim Al-Said, Mrs Dolores Rodriguez Guerrero, Secretary-General of the UN.' Thus introduced Vice-President De Beaufort his guests to each other. 'Please come in.' He led Dolores and Al-Said to a small but cosy room full of country maps. 'Some refreshments will be brought soon. Can I ask the Secretary General why she wanted to have this meeting? Are you really proposing an extensive zwikker indoctrination programme for our parts of the world population? This can't be real?' stuttered De Beaufort. 'Have you become intoxicated by that equipment? Do you believe, Mr Al-Said this is serious?' He almost shouted.

'Can I say something, Mrs. Guerrero?' Al-Said asked.

'Sure, although I hadn't even started yet.'

'Can I also use your first name. Julien, you can't deny our world is transiting a critical era as we have never before experienced. We can't approach that without concern, and some courageous actions have to be taken. I wouldn't be here if I hadn't seriously evaluated the situation. In many African countries people live under the poverty level, and will stay to live that way if they reproduce themselves faster that any economical development can keep up with. Whether this is due to tradition, religion, lower death-rate than before, the fact is there. And all that in the so-called *after-oil era* where primary resources in those countries are limited. This produces tensions as we have recently seen between two countries, where fortunately the conflict had not escalated beyond an initial battle. And why? Thanks to the zwikker and its very courageous team. I'm therefore all in to listen to a proposal from Dolores.'

'Thank you Ibrahim. My first goal is that people must be happy, whether being poor or rich. When getting rid of all indoctrinated complexes, which are and have been always the causes of stress between people, they may be able to take their fate in own hand. This has been shown often in history when due to natural disasters people showed a surprising solidarity. I've always respected as Secretary General all world religions and non-religious opinions, but I really think that a clean-up of dogmas and complexes would serve their people. And unfortunately, this can only be done with the zwikker, not because I like that equipment, but because part of the population is already compensated. A conflict between the compensated and non-compensated should at all prices be prohibited. And that is possible.'

'Stop, Dolores,' interrupted Julien De Beaufort. 'Fighting criminal groups and fundamentalist groups, yes, treating whole populations, no. I've read also the results from Uzbekistan. People have lost their complexes but they build up new ones.

Instincts are not kept under control by traditional patterns and the law. To be honest, you are looking at the world too much as a family. Sorry to say it, but at your position they should never have nominated a woman, how clever she may be. But being as it is, and I suspect you are slightly intoxicated by that zwikker, I will not go further than what I've mentioned. All those people free of complexes seem to have higher reproduction rates than people "having learned their lessons." And in Europe we are not waiting for more immigrants from elsewhere than we can manage.'

'You want me to resign from my post, Julien?'

'That would not be a bad idea in my view, wasn't it that most of the delegates have lost their head and would not be able to nominate a better candidate.'

'Are you a candidate for my post?'

'No, I know my limits, lacking charisma and being a bureaucrat, but I hope you appreciate I'm honest in our discussion.'

'You have the same honest ideas about your own position?'

'Holding your remark as not personal, my answer is yes. I will resign if required, but I don't want anything to do with the zwikker.'

'Lady and gentleman,' interrupted Ibrahim, 'it's great to hear you exchange ideas in a non-diplomatic way, but we need to come to a conclusion. I understand from Julien that he doesn't want any more immigrants from Africa, and that the zwikker might only be applied to people when the law is trespassed. His argument, Europe is already full and mixed up, is naturally relative. He should just promulgate a better birth control among all different people, and therefore the zwikker might help.'

'Perhaps, but not under my presidency.'

O.K., but what about Africa,' continued Ibrahim. 'We have still lots of space for being exploited. Suppose we can develop deserts with modern methods and produce of water, should we than refuse to create colonies from European immigrants? We might certainly need the zwikker for convincing populations.'

The discussion continued for about an hour upon which Dolores prepared a short statement which was co-signed by all three and of which the content was limited to the proposal of Julien. Only zwikker applications by remote control for fighting law abuse when other methods are exhausted. The result was given to the combined group of experts for further elaboration.

Moscow

Dolores travelled subsequently to Moscow.

'How was your meeting with Julien and Ibrahim?' asked Victor when she had arrived in the Kremlin.

'I've learned more about my personal appearance in two minutes from Julien's conversation than I could have done in years of introspective study. But the results, I see you have already on your screen. I suppose you accept that, it isn't too far from your own decision.'

'Indeed. We can talk about that later, since we will go to Rome. I've here an invitation from Pope Paul VIII to join his discussion with the Alberta-group in the Vatican. He would appreciate when you, the head of the Russian Orthodox Church, Yuri Kaspadov and Professor Larson could attend.'

'What's his intention?' asked Dolores when Victor handed over the letter.

'I don't know more than what is in the letter, but the courage and swiftness of action of this Pope is well known. I've already arranged our flights. Our Patriarch can make himself free. Yuri showed interest but wanted Svetlana to come along. If you can get in touch with Professor Larson, we can all be in Rome the day after tomorrow.'

'Well, let me first ring Professor Larson, who appeared in a few minutes later on the display. 'Can you be the day after tomorrow in Rome Professor? Pope Paul VIII wishes a common discussion, with you, Professor Bolotnikov, Yuri, myself and the Alberta-group. Can you be present?'

'Naturally, for such a discussion I can free myself. You want that I present some results?'

'Yes. Do that, but on video.'

'Fine I'll select the most essential components to save time.'

'How are you doing?'

'Well, I feel myself very happy. I wish everybody on the world could have this feeling. You wish I talk about that?'

'Perhaps, perhaps, don't worry about that.'

'I never worry, Mrs Guerrero.'

'Oh, yes, I forgot that,' and saw his cheerful smile zoom away.

'I'm still not used to those compensated. They work on my nerves. Are Yuri and Svetlana the same?'

'No, they have passed the happy-smile-syndrome-period. They have according to me a type of consciousness which we haven't encountered before. Perhaps they will explain when we're in Rome. So far they have released nothing. They talk with each other but are immediately silent when someone comes in their neighbourhood.'

'They don't speak with other persons?'

'No, no, I mean about their experience. Apart of that they act almost normal.'

'But now the next item. We've given it hardly attention. I mean the development of the zwikker for compensation of the gravitation. I've understood that in America they are the point to understand the theoretical context of the functioning of the zwikker on the gravitation. Very soon a request will come to the UN to carry out a large application test. Decreasing the weight of space rockets. The benefit is highest and these rockets are lanced in remote areas. Thus the personnel can be well protected from these super zwikkers.'

'O.K., we'll talk later about it on the airport for the flight to Rome. You can stay here if you like or go to your hotel. Lena will take care of you.'

Dolores chose for her hotel from where she had a splendid view on the golden cupolas of the Abbey in the Kremlin around which flew piebald crows. As long as these will continue to be there, the Kremlin will exist, she had heard. The greenness of the trees was here less green than in Strasbourg. The winter had been very long and people were still wearing a bonnet.

The next morning she was awoken by the telephone. It was Lena who said that she would be picked up in an hour. She had to hurry to be ready. She caught her trunks, made herself up and rang for the bill to be prepared, when there was a knock on the door.

It was Svetlana which stood there. Dolores embraced her and let in her. 'I come to get you, Mrs Guerrero, Yuri is downstairs.'

'How are you, Svetlana?'

'Well, complete well I can say, although I can also say that it doesn't matter.'

'What a strange answer,' thought Dolores. Victor had already said that Yuri and Svetlana had become strange. Incomprehensible in their judgements. 'Just wait a moment, I'll have to check whether I've left nothing behind. A habit, which saved me many times from loosing my credit card.' She found nothing under the bed and in the bathroom. Meanwhile the clerk had come and charged the trunks on a caddie. With Svetlana she went in the lift, walked to the counter and paid her bill. Yuri entered the lobby and greeted her cordially. He invited her to come to the taxi-magmobile in front of the hotel.

'He has the same allure on his face as Svetlana,' thought Dolores. 'But, if I look twice it has disappeared. What is with them?'

Yuri sat in front and Dolores with Svetlana in the back. There developed a conversation about the city and Astrid. Svetlana had recently contact with Astrid.

'What did you discuss?' asked Dolores.

'Astrid had been rather worried. I imagine that Astrid wanted to know more, but Yuri and I are not yet up to that point that we can tell what exactly has happened. I suspect you're interested too?'

'Everything needs its time, also your feelings and thoughts. I can understand what you have done and in particular how you have reacted. In any case you've saved Yuri. Of what, is still the question, but probably of something terrible.'

'I'm glad you take it this way. Finally I've allowed Yuri to compensate a part of humanity. We know both what that means. Yuri and I know even much more, but we don't know yet how we can explain. We've spoken together long about it, or in fact thought more than spoken, because we think the same.'

Dolores saw the haze gliding again over Svetlana's face. She seemed to go far astray, as being absent. Because the taxi magmobile slowed down, Svetlana awoke abruptly and the haze disappeared.

'The airport ladies and gentleman,' called the driver. 'I'll help you with your luggage.'

'Excellent,' replied Yuri, who opened the door for Dolores and Svetlana. The driver delivered their luggage to a porter, whereupon Yuri settled up.

In the hall they met Professor Bolotnikov and the Patriarch of the Russian Orthodox Church. Victor presented Dolores, Svetlana and Yuri. The Patriarch looked for Dolores a very ordinary man. In spite of the worldwide threetel retransmissions she had never met a Patriarch in civilian.

'Alexei Savostin, at your service,' he said. He had deep blue eyes, a short grey beard and moustache and had the same bonnet which carried almost every Russian. Only by the cross one could know that he was a clerical.

'Monsignor Savostin, nice to meet you. I've understood that we all go for a discussion to Rome. But one thing I would be glad to know. Did you saw that special threetel retransmission of the UN?'

'No, I haven't seen it. I am not compensated, such as some of my popes. I hope that our discussion in Rome will help me finding ways how to deal with them. As Patriarch I'm really inexperienced in this matter and there is nothing in the holy books for support. I've understood that you have held some consultations with a religious group. I must once more examine why we didn't participate.'

'You were invited or at least the Russian Orthodox Church. But now you get the chance of making up the delay. The Pope himself is not compensated. That I can assure you.'

'Interesting. Fortunately Victor has given me some information. What concerns the technical side of the zwikker, I mean. Concerning religious aspects, he couldn't tell me very much. You know more...?'

'Not much more, but Pope Paul VIII has invited Professor Larson to report on his recent findings. But I see that we have to check in.'

Dolores got a seat between Svetlana and Yuri, whereas the Patriarch and Victor got seats more in the back of the plane.

During the flight Dolores saw Yuri staring through the window. He seemed to be far away with his spirit. Dolores felt his tension going to Svetlana, who was left of her. She got the feeling as if they communicated with each other straight through her head about something of which she was entirely strange. She allowed herself to drift along and she got slowly an idea what had happened with Svetlana and Yuri. It touched the roots of their existence, whatever that may be. Dolores stared at something that came closer and closer, until she startled with a yell from her trance.

A steward had touched her arm. 'Please don't be afraid! Can I serve you something before diner? A light cocktail perhaps?'

Dolores required some time to recover. 'Thank you, a Brazilian cocktail, please. Such a green one.'

Victor approached from behind. 'What was the matter? You were yelling very loud.' What he didn't say "just like Svetlana when we touched her in her hotel".

Dolores turned around, and said: 'Nothing, I believe. I was dreaming and was shocked when the steward touched my arm. A pity, I would have liked to know the end of my dream.'

The steward brought the cocktail and Dolores sipped from the green fluid. She recovered completely and felt slightly tipsy from the alcohol. But that felt deliciously. When the meal was served, the three distinguished themselves in nothing from Victor and the Patriarch who had a busy conversation.

Vatican

Pope Paul VIII received his guests in one of the restored 15th century apartments. He had chosen this, because he had experienced that people became relaxed. He embraced the Patriarch and greeted next Dolores, Svetlana, Yuri and Professor Bolotnikov. The members of the Alberta-group, Professor Larson and some religious assistants were already present. These stood up and greeted the new comers.

Pope Paul VIII greeted them with the words: 'I'm calling you a very cordial welcome. I appreciate it particular that you all accepted my invitation on such short term. We've something to discuss that concerns the whole humanity. As you know our subject concerns the zwikker. With your authorisation I propose to split up our discussion in two. The first subject concerns the question: "What has been discovered that is of mental value for humanity?" With mental value I mean everything related to the human spirit. As a second subject I would gladly reach a common point of view, or if this is not possible, a collection of points of view, how we recommend a further use of the zwikker.'

The faces of the participants showed several emotions. There were a couple persons with cheerful smiles. Dolores suspected they were from compensated persons. 'Although,' she thought, 'you never know with the clergy.'

'I adopt,' continued the Pope, 'that all of you are informed by the press. What is lacking is information on the result of the study group in New York, which has worked recently on religious meetings. Professor Larson can report us about that. He is here w th authorisation of the UN, isn't it Mrs Guerrero?'

Dolores nodded. She did not say that the UN would have authorized anything.

'But first I have some questions to Professor Larson. You have been compensated, I believe?'

'Yes Holy Father. Fortunately I must add.'

'And how do you experience that, with respect to your original Lutheran education? You see, I'm well informed.'

'Happy is my answer to your first question. Concerning my religion I feel completely relaxed. It doesn't really matters.'

'Must I understand that you lost fate?'

'Not really, I gained fate in a kind of truth which goes far beyond boundaries. Fate like children have who are open to anything and not yet indoctrinated by a heavy load of education, either religious or not.'

'And your opinion about the other religions, of which we have here the representatives?'

'As long people are happy with it, fine, but are they certain they know the truth? Isn't Jesus saying let the children come to me? Children are no theologians.'

'And you still believe in God?'

'As I told you already, it doesn't matter. When I have contact with some of my ancestors, they just say: "We're still there" There is no mentioning of any super spirit which should be the Lord, named Javeh, God or Allah.'

'You still attend the Sunday messes?'

'Sometimes; it's most interesting the common fate of those people.'

'Not your fate, I presume?'

'Not more me, I've the happy feeling of an individual, not being bossed by any person or system.'

'You are quite franc, Professor Larson, but I suppose you are honestly expressing yourself.'

'Holy Father, you haven't asked for lies.

'Than a last question. How does Jessie Evans, the mongoloid boy which you've treated?'

'Well. He gained consciousness which makes him a happy child. His brains reacted positively to the treatment by the zwikker, which is repeated every three months.'

'I am happy to learn that, but let us concentrate on your scientific investigations.'

Professor Larson gave a sign to the technician in the corner of the room. Light weakened and a threetel screen lighted up. On the screen a church became visible. It was as if the witness walked along the people into the church. A Catholic service took place and a priest was busy with the preparations for the consecration. The priest showed first the host and next the chalice. Professor Larson said: 'In the right corner of the screen the spectrum of the priest is projected.'

With full tension they followed the images. Would the spectrum be modified?

'Now you must pay attention,' said Professor Larson. 'You'll see that the spectrum of the priest continues to preserve the same form.'

'The next part of the video concerns the communion. The operation is shown first, then reversed and again accompanied with the spectrum of the people who go for the communion.'

'Look especially on the right part of the spectrum.'

What they saw was a weak vibration. This happened each time when someone took the host to himself. The strength of the vibration differed from person to person.

'So far these selected images from a catholic service. The following pictures concern recordings in a Mosque.'

Dolores tried to see whether the Muslims under the present guests would react different from the Christians and non-religious. A lady dressed with head scarf stood up to get a better look.

The spectrum differed hardly from the previous experiment. Automatically they looked at the right hand part of the spectrum whether this would start to vibrate. When the service lasted longer one could observe indeed the same vibration such as those at the communion.

Pope Paul VIII stood up because he thought it was the moment for a discussion, but Professor Larson who saw this, said: 'Before you discuss these results I will show you in quickly some other registered pictures. We've been also with the zwikker at other religious meetings.'

They saw the same vibrations during a Hindustani worshipping, at a Buddhist pilgrimage and during a ritual Indian dance. The conformity was striking.

Professor Larson gave a sign and the technician switched off the video.

Everybody blinked with their eyes, slightly upset. Dolores was curious how the clergy would react.

Pope Paul VIII took the word. 'In our innocence or guilt, call it as you wish, we as a clergy think that we have the truth concerning God and our relation to God. From the fact that we are here present as a multi clergy, it becomes clear that there are several ways of people to God. That does not prove, however, that there are several ways of God to people. If the last observation of Professor Larson is correct, the zwikker measures the same consciousness of people if they address themselves to God. It does not matter how.'

'Their consciousness is in vibration, I rather would say,' interrupted the Patriarch.

'And what I observe,' spoke the Humanist, 'the zwikker cannot measure therefore God. At least not the presence of God during a service.'

'That depends,' spoke the Lutheran Eckelhof, 'the vibrating shows that people feel something special. And therefore God exists for them.'

'It is irrelevant how you try to describe it,' sounded it suddenly from the corner where Yuri was seated. He had stood up. His face was pale.

Victor and Dolores looked at him anxiously. 'Would Yuri have experienced something that could give an answer?'

'Svetlana, my spouse,' Yuri took Svetlana's hand and let her also stand up, 'Svetlana and I, we know that you're all at the same time right and wrong. I will tell you what we have experienced and explain why.' A gentle haze appeared on both faces.

'When Svetlana aimed the sensor on me and held me tight, we were both not only compensated but pushed through a threshold we've never experienced before.' He brushed with his hand over his forehead as to remember the moment it happened. 'We became both unconscious. Our spirit seemed to mix itself and was carried away in time and space. One can also say without time or space. There was something else where no dimension counts. This situation seemed to last indefinitely and nevertheless it was but a moment. Where we were, we didn't know. There was no question of a certain place. It could be at the same time a point or an infinite space. We discovered that spirit exists both apart from substance and is also linked to it. In our body is spirit with a bridge to substance. But where we were, our spirit was in connection with something omnipresent. It has to do with something very ancient which was there before the existence of the world and the universe. But it's here now, too. We didn't feel ourselves separated. We felt us linked with each other and with all other beings. With that I mean not only humarity, but everything that lives.'

Yuri stopped; there was some colour on his face.

'There is more,' interrupted Svetlana, 'there was a consciousness of billion of years. It gave me a chock when I had to depart from that. I have screamed, they told me when I returned from my unconsciousness. Yuri and I have - thus we think – again a consciousness spectrum, but then one that coincides with the source we visited. We've understood that the zwikke carried us back to that source. It has much in common with the experiences of people who returned from an apparent death or coma.'

Svetlana sat down again. She still held Yuri's hand. Nobody dared to interrupt Yuri or Svetlana.

'For this reason', continued Yuri, 'you are all almost right, what means just as much as almost wrong. There is more than we think or is told by the religions. That source is constantly fed by us, you and me. Moreover it evolves with the creation and lasts by means of an evo utionary programming of births and life. It's only due to our lack of knowledge that you interpret this source all differently. As a result, it does not matter how you do that.'

The words "it does not matter" continued to resonate in the room.

Dolores was deeply moved by the story of these two particular persons. Would it be possible that their message was understood? Hadn't earlier people told the same tales and were taken as fools?Pope Paul VIII stood up. 'Mr Kaspadov, you have given here a deposition which should invite us to deep meditation and reflexion. You and your spouse have had an experience which we hardly can imagine. Would you like to take part in the further discussions? How do you think about that?'

'Holy Father, Svetlana and myself will prefer to withdraw. You'll understand that we don't see any necessity of your discussions and we aren't entirely stabilised physically. We ask your permission to withdraw.'

Yuri caught Svetlana by her hand and left the room.

After their departure there went a hullabaloo through the room. The Pope offered to pause and drink coffee or tea in the adjacent refectory. Everyone accepted gladly to recover from the undergone emotions.

'That was a beautiful performance,' said Jitah Crita, the Lamaist to Venu Krishnamurthy the Hindustani. 'Those westerlings understand at last. Something we know already for a long time. Those two people happened to be very persuading.'

'You want also coffee, dear?' asked John Wiley, the Anglican archbishop to his wife.

'Thank you John, and you know I believe those two youngsters. I wouldn't dare to participate in such an experiment, but it shows that our faith has just as large a value as that of the others.

'What we anyway should avoid is to argue on differences between religions. We must look for the core and it does not matter how, by means of a religion or without religion.'

'But John, if you proclaim that at home we're soon on the street!'

'That's the problem of humanity. Everything is located in boxes and structures and of our spirit we can't live, how interesting that source of Yuri Kaspadov is.'

'What I found particularly interesting, John, was that remark on all living beings. I've always said to you that our cat has a soul and our chickens have more in them than we think. Look how happy they can be.'

'Oh, you with your cat and chickens.'

'Indeed, and I will bring that forward later on. Perhaps is there one who agrees. That Hindustani perhaps, what's his name again?'

'Krishnamurthy, but then you must talk about cows.'

'Then the story of that female Doswon about a living emptiness in atoms and the universe has indeed some sense,' she mumbled. 'I'll have to study that more thoroughly.

The meeting was shortly resumed where the initial discussions resulted mainly into views that all zwikkers should be banned from the surface of the earth, because a new wave of compensation would not let the religions survive. However, a lively plea of Professor Larson on its usefulness made a big impression in particular with his successes with Mongoloid children. The case of the boy Jessie Everton had made great impression in medical circles. After a long discussion they agreed that the zwikker should only be applied very restrictively. This provided that religious people would be spared of such treatments.

Rome

The next day was full of sunshine and Yuri, Svetlana, Dolores and Victor had tea on a terrace in Rome. They overlooked the old ruins of the Roman Empire. They were tired, all four. The meeting had rather used them up.

'Actually I don't know what I had hoped,' said Dolores. 'I had expected a much more thorough discussion after the intervention by Yuri and Svetlana. Science and faith can complete each other, and one could at least have confirmed that. But, on the other hand some kind of understanding is born although I've the impression that understanding their own religion is the largest problem.'

The others did not react.

'You know what is always on my mind? I wonder whether it has sense to continue as a Secretary-General of the UN. How you think about that?' aiming at Victor.

'I'm just looking around and try to understand the history of empires. Why people cannot simply live next to each without any State organisation. Eh, what did you say, you want to resign? Do I hear that well?'

'Eh yes, I'm a bit tired, and sometimes I desire to live an ordinary sociable life with Jorge.'

'Perhaps you aren't wrong, but don't do it now. Take the decision when the UN has recovered. I mean when your delegation members are again in disagreement with each other. Moreover Yuri have said that it does not matter.'

Yuri smiled. 'When can I possibly resume my work? I feel myself since yesterday delivered from a burden. I want to keep my invention manageable, without risks of addiction.'

'Good, that's what I like to hear. You have lost that haze around you. That applies also to Svetlana.' It was Dolores, who had said this. She aimed at Svetlana. 'Svetlana, what are you dreaming about?'

'I'm thinking of the little son that I will get. He has received something of the source, I'm convinced of that.'

Before one got the occasion to congratulate, Professor Larson approached loudly calling. 'Youngsters, can I join you? What is the matter? Why you're shouting with joy?'

Dolores saw that his cheerful smile had been replaced by his sympathetic expression of the former days which made this Scandinavian so sympatric with his people.

21

If spirit is everywhere and we are part of it, where is our responsibility?

New York

It lasted still two years before the UN gave its fiat for a further development of the zwikker. The disaccord between compensated and non-compensated people had entirely disrupted some States and had caused a number of civil wars. The UN had not been able to intervene everywhere. Generally this came because non-compensated had seized power. The destiny of the pleasant compensated had as a result deteriorated, and it had lasted years before they managed to cope with it.

Dolores was still Secretary-General of the UN. She was in her office. She read the last report on the zwikker developments and waited for two visitors who had asked urgently for an audience. Since their names were Mr and Mrs Troover from her birth country Chile, she had become curious. The name Troover intrigued her. Would it be Peasley? So many Troovers did not exist in South-America, and he had gone up in smoke, in spite of all investigations.

The request had come by her private number, and few people knew this number. A sympathetic woman voice had addressed her in Spanish. She had given no characteristics, but she had insured that all was entirely safe. Dolores had to fear nothing. For certainty she had warned the Interpol and two of their agents were present in the adjacent office. She could call them by pressing a button.

She startled unnecessarily when the intercom announced the visit. A bit nervous she was nevertheless. 'Mr Benjamin Troover and Mrs Gloria Troover for you.'

'Let them come in.'

The door opened and a nice woman entered, followed by a heavy man. Dolores forgot to greet them. She had never met Peasley in person. No doubt, he was it. With a small beard and a large moustache. His hair had to be a wig. When Peasley took off his sunglasses it was even more clear.

'Eh..., cordially welcome. You are Mrs and Mr Troover?' she asked superfluously.

'Indeed. It's very kind of you to receive us.'

'Is your name really Troover? Or I can say Peasley?'

'You do recognise me? I'm indeed Peasley, but under the name of Troover we have married. I would appreciate if you wouldn't call me Peasley.'

'As you like, Mr Troover. What's the reason of your visit? You know you're still a wanted person? I can let you arrest within two minutes.'

'I've assessed that risk, I know the world. I request you to listen first to me. Afterwards you can decide whether you still want to let me take into custody.'

'That's reasonable. Finally I've already received you. But before you start, have a seat and can I offer you something to drink? Otherwise I feel myself a bad hostess.'

'I'm no longer the former person who disappeared with a zwikker to Brazil, which by one of your ladies, thus I call them, was destroyed. Nice work. After my escape from Montevideo I took refuge in Chile. There I've seen that unfortunate UN retransmission. I've afterwards understood that my subconsciousness was wiped out, except some complexes. I had probably more complexes than the average man.'

'Ah.., for this reason you happen to be different than I had expected. You were rather a grumbler in your former position.'

'My insights have not changed, however. I still believe that people are less to be trusted when being higher in the society, compensated or not.'

'And that you say to me, who has one of the highest posts in the world?'

'I've not come to flatter you. As former head of US Security Service I know that one will do things which ordinary people would not do on ethical grounds.'

'I see you've recovered some your former consciousness. So, you've come to offend me?'

'Certainly not...! I try to be only honest. At your level it has no sense to tell lies.'

'I've heard that saying before. My spouse thinks also this way. As a result, I'm perhaps somewhat less "unreliable" as you call that. With that I flatter myself. But why do you tell me this all? I still don't understand your intention.'

'I've still an old zwikker.'

'What do you say!! Is that true? Yours had been nevertheless destroyed!!'

'That lady of you has indeed taken along the destroyed zwikker. But not the copy, which I had made in Niterói. This one is still in a safe in a bank in Rio de Janeiro.'

'Ah, for this reason the discussion concerning the scrap pieces of the zwikker had been closed. They confirmed in Moscow and Washington that the pieces were from an original zwikker. And have you ever used that in Rio?'

'No. However, I intended to recover it later when Interpol would be less in search of me. But after that compensation I wasn't more interested. Moreover I married with Gloria and we had a busy life in agriculture.'

'But what do you want from me actually? That's still not clear Mr Troover!'

'As you have already noted I've recovered some of my old consciousness and I want to work like in former days. With or without zwikker. Someone must stop me using my zwikker. You are that someone. I can become employed by the Chilean Government who gladly wants to use my capacities.'

'I still don't understand why me?'

'By accepting that offer of the Chilean Government I run the risk of being unmasked and taken into custody. I can of course prevent this by picking up my zwikker and with that handle everything... But Gloria has begged me not to do that. For this reason I propose an exchange: You get the key of the safe in Rio and I get complete amnesty. You have the right to do that on the basis of the statutes of Interpol which falls under the UN.'

'Is there still more?'

'Yes, you obtain a copy of this report in which I give my honest opinion on the developments of the last two years. You cannot deny that it has been a mess. Here I have with me the amnesty form. You have only to sign.'

'How did you get that? You have burgled our office?'

'No, that wasn't necessary. You know very well that your system is as leak as a sieve? Do you still trust your delegation members and employees? You're not that naïf, I suppose. I came through by means of my computer, without any problem. I still remembered the entrance code. When I was in service at US Security Service I could tap each telephone conversation, code or no code. Also that of all important statesmen in the world. And since it has only become easier.'

'I'm always taking that into account.'

'For this reason the Chilean Government wants to make use of my talents.'

'Mr Troover, you let me hesitate to allow you to walk free around. I cannot determine whether you are a danger for the community or not.'

'Let I remain honest. You know better than who else, what moves people to predominate others. With that, complexes play a role. Some do that by revenge, others by stupidity. You have seen that entirely compensated people have no desire to dominate others. I myself had in former days a gigantic complex. I refused to be ever laughed at. For this reason I had to be the boss, under whatever circumstances. If this is the case for other high officials and politicians, then they are harlequins of their complexes. There are few who act from pure idealism to commit themselves for others? And look at the people around you. And, if I still may continue, all difficulties which the world has experienced since the first zwikker have rather been caused by ignorance than by statesmanship. I suppose you can't deny that, do you? And are you yourself satisfied with what you have done?'

'On that I haven't to give account to you Mister! Moreover, a high official must do it with his own qualities, knowledge and intuition. He can't rely on any employee. If he asks a confidential advice, the employees build always something of self-interest in it. The official stands therefore alone, complex or no complex. As a result, I say, Mr Troover by standing alone, you failed some years ago.'

'Perhaps you're right. I repeat, however, my question. Are you satisfied with the results of your career?'

'Do you think, Mr Troover, I better say Mr. Peasley that something can be improved?'

'Yes, and initiated by the UN, how difficult that may be. And with the newest zwikker. Responsible persons must be tested not only on their knowledge but also on their complexes. My suggestion is to compensate them temporarily. Subsequently they can recover with parts of their subconsciousness. They know then at least how to handle themselves.'

'Something, as what you have experienced yourself?'

'Not really. I'm not yet to that point that I want to recover parts of my consciousness spectrum which are stored in my zwikker in Rio. Gloria is against that. She knows me as compensated and wants nothing else.'

'Is that true Mrs?'

'Yes, certainly. It's even a condition on my part. Benjamin has told me that he was in former days shy of women, and that is the last thing I wish. At least towards me.'

'Do you think Mr Troover that the world would improve if governed this way?'

'There would be anyhow fewer conflicts. And then you've gained already much. Conflicts arise always from old revenges. That happens on small scale in families and on large scale between regions or states. And when you have a couple of extra complexes a conflict is born.'

'You resemble a philosopher who has been for years in meditation. This I hadn't thought of you, honestly.'

'It is not wisdom, it is only based a simple and healthy distrust.'

'What do you think I must do?'

'Resign, you are already too long in service. That is always wrong. Habits become truths, whereas your tales become lies, how well meant. Moreover you stand weak because of all problems in which you've been involved.'

'You dare, Mr Troover. You dare. You apply that on yourself, too?'

'Probably not. But Gloria is fortunately at my side.'

'What resigning concerns, I'll do that accidentally over a month. The notice appears the day after tomorrow. What is there in your report?'

'Exactly what I've told you, but then confirmed with practice examples. I've analysed a number of conflicts and given my opinion. It's in the first place intended as a support for the Secretary-General of the UN, so your successor.'

'This was a remarkable conversation, I must say. I must think about it. Can I contact you later?'

'No...! I leave from here with a signed amnesty form, or you see me disappearing in a different way on which I will not comment. Your two guards in the adjacent room cannot stop me.'

'How do you know that? You force me to take now a decision....?'

'Indeed. You are not a person who is afraid of responsibility. I've been honest to you. My zwikker is at your disposal to be destroyed. If I don't report tomorrow to Rio, the zwikker will be given to a designed person.'

'I'm not afraid about your threats. You can do nothing with your zwikker? As soon as you switch it on, it is destroyed automatica ly.'

'No. The last years I have studied all facets of the detection and destruction cf wild zwikkers. But that doesn't concern zwikkers which are in a cave and surrounded by an electronic cage.'

'What can you do with a zwikker if it's in a cave?'

'It aims at threetel systems which transmit directly. Thus I can reach the complete world. But..., I don't want something like that, as long as I feel myself much better than in former days. You should compensate yourself. It feels very relaxing.'

'Although I judge your presentation as blackmail, Mr Troover, I must recognise that it would be better if there were more people on the world such as you, at least such as you are now. I've encountered many statesmen in my career. I rather appreciate your honesty and will forget the blackmail tendency. I'll sign the amnesty form.'

Dolores caught the form, read it completely and signed it. She added a note that this amnesty applied Tim Peasley, alias Benjamin Troover.

'Thank you very much Mrs Guerrero,' said Gloria. 'I've heard much about you, but you are still braver then I had assessed. Benjamin should not have threatened you of course, but he believes that someone at a high position must know all the facts. He does not consider t as threatening. Here is the key of the safe in Rio, including the address and an authorisation. The other key will be destroyed next month. You'll allow us now to leave?'

'Yes, and if I visit later Chile I hope to see you both again. This conversation deserves a continuation, I think. But not in my position as Secretary-General. Take care.'

When the two visitors had disappeared Dolores remained long silent. Hadn't she been taken nevertheless by surprise? She startled from a heavy noise and tapping from the adjacent room. She found there two Interpol agents tied up to their chairs. She rescued them, upon which they leaped up and called incensed: 'Where's that bear who overpowered us. He came in, gripped us and before we knew we were fixed in ropes.'

She caught a sheet of paper from the floor which had been obviously written in haste. There stood on: "Our excuses for this special treatment. It was necessary to ensure our safety in case you hadn't signed. We wish you the very best. Signed B. & G. T."

Dolores bursted out into laugher in front of the bewildered faces of the agents. 'Gentlemen, ha, ha..., you've had the honour to be tied up by the former head of US Security Service, Tim Peasley, alias Benjamin Troover. Don't take it too hard.'

'But Mrs, against this person runs an international arrest command?'

'That has been raised. In my function as Secretary-General I have the right to grant amnesty. I will inform the head of the Interpol of my decision. Come in my office and have a drink. You've involuntarily assisted to solve a difficult question.'

When the men eventually left, Dolores searched for the last report that concerned the zwikker. 'Look. Only two cases have been on unlawful "in"doctrination". The apparatus had been traced within seconds and had been destroyed. It wasn't mentioned whether the operator had perished thereby.'

Her feelings floated between joy and desolation at the anticipation to take farewell in a month. Peasley was right, she should be no longer Secretary-General. Jorge called it her "camel" syndrome, as camels show from the moment they enter a zoo. They walk after years the same circles.

But "camel" or not, she nevertheless felt it. Fortunately she would be succeeded by the Chinese Lin Yang, who had attended the first discussions concerning the zwikker in the UN. Lin had shown wisdom and could possibly solve the current problems better than she herself. She would give him Peasley's report.

Dolores intended to ring Victor for the last time. Victor had never run out on her these last years. He had been reappointed and would be a support for Lin Yang. What concerned the others, Julien De Beaufort was oddly enough still in function. President John Smith of the US had been succeeded by a party member, called Levelyn Darsy. Darsy's policy concerning the zwikker had been especially aimed on space travel and the development of super-zwikkers. Each super-zwikker could bring thousand kilos of mass in space. The objective was to improve the already functioning of energy radiation to earth stations.

It was already after midnight when she got Victor on the line. It was early morning in Moscow. Victor looked fresh and energetic with a twinkle in his eyes.

'Are you attending my farewell?'

'Sure.'

'I have a couple of questions.'

'I've half an hour is that sufficient?'

'Yes, enough. First of all something about Peasley. He visited me this morning.'

'What! He came to your office?'

'Yes something like that. His wife had asked for an emergency audience. I have granted him amnesty.'

'Just on your own authority?'

'Yes Victor, on my own authority as head of Interpol. There's more in that man then we have thought. I will inform you later. He is now free man. Please pass that to your services. You will see him one of these days as a Chilean official.'

'I continue to be astonished with you! Sorry you will resign soon. You were the salt in the UN pulp. I'll miss you. What do you want to know further?'

'How are Yuri and Svetlana and their son?'

'Yuri and Svetlana have put everything in a book, but that hasn't been made public. They doubt if their knowledge is appreciated by everyone and they want anyhow no attention of the world press, in particular not about their son. He's very special. Much of their experience has been confirmed by volunteers who had been treated identically. I'll send you some of the results. Is Pope Paul VIII coming also?'

'Yes, as a Head of State. His seat will be close to yours, just like President Darsy. The places have been classified alphabetically. Victor, once again thank for all you did for me personally, and see you soon.'

UN building

Dolores stood for one of the windows of the translator rooms and saw people streaming into the Assembly Hall. She saw Julien De Beaufort pottering around to find his seat.

'Only this afternoon, then I'm free,' she thought. She left the translator room and returned to her office. Hermann Janitsch, who would resign next year as President, would introduce her.

In her office she saw Jorge, their two children Natasha and Robert and Soraja drinking tea. Jorge and the children were lively. Soraja had tears in her eyes.

'That's sweet of her.' Whereas we shared all confidentialities she has never become really close to her. She had remained always the perfect secretary.

Jorge made her a cup of tea and asked: 'Do you have your speech at hand?'

She startled. Of course she knew that the papers were in her bag, but for certainty she checked it. Jorge grinned. 'Come on, if you can't handle this, nobody can.'

Jorge wanted to continue, but Hermann Janitsch came in and said: 'Ladies and gentlemen, I request you to follow me. The Hall is waiting for us.'

All, except Soraja, stood up. Soraja had refused to go to the hall. She did not find it appropriate for a secretary. In her heart she preferred to follow everything on the screen.

Since the zwikker accident with Yuri, there were no more direct retransmissions of political events. One considered the chance on an accident too large.

Dolores, Jorge, Natasha and Robert kept track of Hermann. Near the hall she heard people talking by greeting each other and exchanging news. All stood up when Dolores entered the podium.

Hermann Janitsch, who always tapped first on the microphone, started immediately: 'Excellencies, Heads of States, representatives of the world religions, dear delegates and ladies and gentlemen before their threetel, welcome. Today our Secretary-General resigns and she has expressed to be the only person to speak. Mrs Guerrero appreciates, however, to meet you at the reception.'

A whispering passed through the hall. Not everyone had agreed with the initiatives of the Secretary-General and again she showed her own determination.

Dolores caught her bag, gave in passing the President a hand and walked to the pulpit.

'Nobody is perfect, Mr. Peasley told me this early this week.' She paused a moment and said then: 'Almost right is the same as almost wrong! Excellencies, representatives of world religions, delegates, and all in the world who listen to me.' She thought a moment: 'How's my hair,' and continued, 'My service period as a Secretary-General has been characterised by an apparatus which was developed to measure and adjust the consciousness and subconsciousness of people. We have debated here in this hall many of its consequences. I've taken decisions in deliberation with Heads of State and alone. Not all was a success. The boss of the US Security Service was fired, an American habit, with serious consequences. He escaped with a zwikker. However, indirectly through him the danger of addiction to the zwikker was discovered. Unfortunately too late, so that a quarter of the world population was compensated. The UN and a lot of States have experienced the consequences, economic statistics showed ups and downs, more babies were born and in general many persons gained in happiness. Peasley was never found, but entered last week my office.' A murmuring passed through the hall. Two people stood up and sat down again.

'I've given Peasley amnesty. My arguments you can find in the report on your desk.' People started to open it and trying to read it in the half darkness of the Hall.

'You can read it later on, I suggest. I'll talk of the good things of the last years. Thanks to the zwikker, incurable mental illnesses have been cured; everyone may have read about Jessy Everton, a mongoloid boy who almost became normal. Wars have been stopped and prohibited, some to the risk of the people using zwikkers at the battlefield. I'm proud of many UN officials who helped putting a new order on the rails, with less bureaucracy, less harassment, more positive spirit and less poverty. They deserve your admiration.'

People applauded vaguely. Some delegates wondered themselves what she had in mind. They were still shocked by the words about Peasley and bureaucracy.

'On the other side,' continued Dolores, 'religions were forced to seek the truth, not really appreciated by those who feared loss of fate. It has never been my objective to avoid discussions, in particular not when based on honest opinions and respect. This respect should concern always individuals. We may not always react that way, but doesn't everyone hope on happiness in some way or another?'

'We have here as United Nations a responsibility to our descendants. Time should once come that good is earlier inherited than bad. Our subconsciousness can play thereby a role. It are not only our genes which we pass on, but also our spirit coupled to consciousness. Is that new? According to me not. The religions try that already for centuries, so why shouldn't we as world population.'

'What has this to do with the United Nations of which I take farewell today? Nothing...? Or everything....? If we want to improve live, we should not behave as ostriches which put their head in sand. The United Nations has had always the objective to improve relations between people. As you are here present tocay, you represent a large number of state forms, each with its own system of governing board. Do we do have to strive to one type of state form, being the bests for all? Within certain limits perhaps? "It doesn't matter" as long as this form respects the rights of the individual. But rights can also be interpreted differently. Human rights aren't always matching with individual freedom of movement and security. In particular where safety of individuals and children is concerned.'

'At present there are five hundred operational zwikkers in the world and more will follow. This apparatus might help us. The UN has regulated the general use of it and apart from some exceptions this seems to be fruitful. Statistics show a decrease in the number of crimes. Also young people, who have problems of identity, can profit. Suicides of young people have drastically decreased. This development is hopeful and I think that humanity can benefit from the apparatus.'

'The task for the clergy is not simple, but she might overcome that by searching for the basic root of their faith.

When scriptures are not sufficient, each individual has inherited a brain to be used.'

'My last words go out to all people in this world whom listen to me. Our existence is more than what we see and feel daily. The fact, that prehistoric people sensed this already, is a proof. We don't have to be frightened of these developments. Humanity in the best sense, should always be our goal.'

'The nature of my speech is that people should constantly wonder if he or she wants to live with complexes and a distorted subconsciousness. In other words be itself and not a reflection of a group or a tendency. An essential component for happiness.'

'I will end with a poem:

People want an identity, which links him and distinguishes him.
But doesn't he lose with that himself?

Since all our joys and conflicts are linked to identity, where's our own value?
Why being afraid to be less original and less free?'

The universe is active empty, as also our atoms and molecules.
Our mind has there its source.

What a space in infinity.

'I will now hand over my task to Mr Lin Yang. It was a great honour to have served as Secretary-General of the UN. Thank you.'

It remained quiet in the hall. What a bizarre end. The last words had called for several responses and emotions. Most delegates looked at each other somewhat upset.

Then Pope Paul VIII stood up and applauded. The response was that everyone stood up and Dolores received an ovation such as she never had had before.

Some tears welled up in her eyes and she realised suddenly that this applause affected her profoundly. She caught her bag, took a handkerchief and cleaned her eyes. Then she waved back and laughed to everyone in the hall and thanked them by applauding herself. She maintained that minute-long, returned to the podium, embraced Jorge and her children and left the hall.

The bell sounded to indicate that the meeting had been raised and people left the hall. Some looked at each other, as if they saw each other for the first time. Dolores had carried them away to another dimension. Like after each emotion, people ventilated that with lively laughter and the reception started in excellent mood.

Lots of people cued up to take farewell of her, from a firm hand shake to a look with tears. Patriarch Aleixei embraced Dolores and said: 'I have learned from you in this half hour more than I had hoped to learn ever more. I thank you warmly for that.' There was some twinkling in his eyes when he said: 'If I were you, I would nevertheless undergo a zwikker investigation. Your task is no longer to carry the world, but to live happily with your husband, children and grandchild. To reach that, you must compensate part of your subconsciousness. I wish you a lot of luck.'

Julien De Beaufort descended next to her and gave her starchy a hand. He could not leave behind to say: 'Dolores, you've done much for the UN, but you should have stopped that zwikker business. I'll inform your successor,' whereupon he just walked rigidly away.

'Don't pay too much attention to his words,' she heard speaking in Spanish. 'You've opened more ways than any statesman or mental leader has done so far. As a Pope I should have undertaken something like that, but you've done it better than me. I know you aren't a practitioner catholic, but allow me to bless you as a deeply touched priest.'

Dolores got again tears in her eyes. She couldn't stop it. Jorge who saw it said: 'I'll help you to escape. You don't have to shake hands with everybody. On the other hand many like to talk with each other and with your successor. I see there Lin.'

Jorge took Dolores's hand and walked up to Lin. 'Mr Lin, we want to disappear unnoticed. Would you be so kind to thank everybody on behalf of Dolores?'

'I'm glad about your speech, Dolores. That will facilitate my task. You've opened many doors and I will ensure that they will not be closed again. Come along with me, we simulate to discuss something and then you disappear quietly.'

Lin guided them to an exit, greeting people in the meantime. There he gave Dolores a hand and bowed. 'You've touched the perpetuity in my spirit, thank you again for all,' and withdrew.

'Now, my girl you've brought it good to an end. Let us go rapidly home. We've a surprise for you.' Jorge caught her hand and Natasha took the other. With Robert behind them they ran to the head entrance where a magmobile stood ready. To Dolores's surprise Yuri was the driver and Svetlana and Astrid sat in the back of the car.

'Oh, how nice to see you! You were also in the hall?'

'No, we followed everything with Soraja. She arranged that and we will bring you home. A diner is already prepared, and Svetlana will play for you.' Like usually, Astrid had spoken.

'Svetlana, did you bring your son Andreï with you? Victor told me he is very particular.'

'Yes, my mother is with him in our hotel Tomorrow, if you want you can see him, and if I'm not mistaken he will ask questions, since he has followed the emission of your farewell speech.'

'In which respect is Andreï particular? Your were pregnant during that zwikker accident, wasn't it?'

'We don't yet know everything. Andreï is a normal intelligent boy, but sometimes he reacts as if he carries with him a wisdom of centuries. That comes up at the craziest moments. If he looks at you, you have the feeling he looks deeply in you. Yuri and I think that he has got as an embryo something of the "source". Generally he penetrates as a two years boy directly to the core. For himself everything is completely normal. He certainly likes to meet you.'

Jorge interrupted with: 'Svetlana, can I have Dolores for a moment? Natasha requires help in the kitchen. She cannot find something and I don't know either where it is. Soon you can babble further on.'

'Oh, how I'm feeling happy!' exclaimed Dolores. 'I felt rather nervous for this day. It's sometimes difficult to perform at a high level. And now you take me along to the every day life. And that's so very real. Thank you, thank you.'

She swept a tear from her eye and ran with Jorge to the kitchen.

==============

www.ingramcontent.com/pod-product-compliance
Lightning Source LLC
Chambersburg PA
CBHW051440170526
45166CB00001B/54